Book 7

Mechanisms of Hormone Action

REPRODUCTION IN MAMMALS

Book 7

Mechanisms of Hormone Action

EDITED BY

C. R. AUSTIN

Fellow of Fitzwilliam College
Charles Darwin Professor of Animal Embryology
University of Cambridge

AND

R. V. SHORT, FRS

Director of the Medical Research Council
Unit of Reproductive Biology
Honorary Professor
University of Edinburgh

ILLUSTRATIONS BY JOHN R. FULLER

CAMBRIDGE UNIVERSITY PRESS

CAMBRIDGE

LONDON NEW YORK NEW ROCHELLE

MELBOURNE SYDNEY

Published by the Press Syndicate of the University of Cambridge
The Pitt Building, Trumpington Street, Cambridge CB2 1RP
32 East 57th Street, New York, NY 10022, USA
296 Beaconsfield Parade, Middle Park, Melbourne 3206, Australia

First published 1979

Printed in Great Britain at the
University Press, Cambridge

Library of Congress Cataloguing in Publication Data

Main entry under title:

Mechanisms of hormone action.

(Reproduction in mammals; book 7)

Includes bibliographical references and index.

1. Hormones, Sex. 2. Reproduction. I. Austin, Colin Russell, 1914– II. Short, Roger Valentine, 1930– III. Series: Austin, Colin Russell, 1914– Reproduction in mammals; book 7.
QP251.A87 bk. 7 [QP572.S4] 599'.01'6s [599'.01'6]
ISBN 0 521 22945 6 hard covers 79-16287
ISBN 0 521 29737 0 paperback

Contents

Contributors to Book 7

J. R. G. Challis
Department of Obstetrics and Gynaecology,
University of Western Ontario,
London,
Canada.

J. Dorrington,
Charles H. Best Institute,
University of Toronto,
Toronto,
Canada.

A. P. F. Flint,
Institute of Animal Physiology,
Babraham,
Cambridge.

H. F. Frazer,
MRC Unit of Reproductive Biology,
37 Chalmers St,
Edinburgh.

R. B. Heap,
Institute of Animal Physiology,
Babraham,
Cambridge.

E. V. Jensen,
The Ben May Laboratory for Cancer Research,
University of Chicago,
Chicago,
USA.

W. I. P. Mainwaring,
Department of Biochemistry,
University of Leeds,
Leeds.

Books in this series

Book 1. Germ Cells and Fertilization
Primordial germ cells. T. G. Baker
Oogenesis and ovulation. T. G. Baker
Spermatogenesis and the spermatozoa. V. Monesi
Cycles and seasons. R. M. F. S. Sadleir
Fertilization. C. R. Austin

Book 2. Embryonic and Fetal Development
The embryo. Anne McLaren
Sex determination and differentiation. R. V. Short
The fetus and birth. G. C. Liggins
Manipulation of development. R. L. Gardner
Pregnancy losses and birth defects. C. R. Austin

Book 3. Hormones in Reproduction
Reproductive hormones. D. T. Baird
The hypothalamus. B. A. Cross
Role of hormones in sex cycles. R. V. Short
Role of hormones in pregnancy. R. B. Heap
Lactation and its hormonal control. Alfred T. Cowie

Book 4. Reproductive Patterns
Species differences. R. V. Short
Behavioural patterns. J. Herbert
Environmental effects. R. M. F. S. Sadleir
Immunological influences. R. G. Edwards
Aging and reproduction. C. E. Adams

Book 5. Artificial Control of Reproduction
Increasing reproductive potential in farm animals. C. Polge
Limiting human reproductive potential. D. M. Potts
Chemical methods of male contraception. Harold Jackson
Control of human development. R. G. Edwards
Reproduction and human society. R. V. Short
The ethics of manipulating reproduction in man. C. R. Austin

Book 6. The Evolution of Reproduction
The development of sexual reproduction. S. Ohno
Evolution of viviparity in mammals. G. B. Sharman
Selection for reproductive success. P. A. Jewell
The origin of species. R. V. Short
Specialization of gametes. C. R. Austin

Book 7. Mechanisms of Hormone Action
Releasing hormones. H. F. Fraser
Pituitary and placental hormones. J. Dorrington
Prostaglandins. J. R. G. Challis
The androgens. W. I. P. Mainwaring
The oestrogens. E. V. Jensen
Progesterone. H. B. Heap and A. P. F. Flint

Preface

Reproduction in Mammals is intended to meet the needs of undergraduates reading zoology, biology, biochemistry, physiology, medicine, veterinary science and agriculture, and as a source of information for advanced students and research workers. It is published as a series of small text books dealing with all major aspects of mammalian reproduction. Each of the component books is designed to cover independently fairly distinct subdivisions of the subject, so that readers can select texts relevant to their particular interests and needs, if reluctant to purchase the whole work. The contents lists of all the books are set out on the previous page.

We have taken the opportunity in this, our latest addition to the series, to explore in depth the mechanisms of action of the reproductive hormones. This subject was only touched on briefly in the preceding books, and yet hormone action has attracted an enormous amount of attention in recent years. It is a field in which the biologist is often unaware of the biochemical advances, and the biochemist is frequently ignorant of the basic biological principles. We have attempted to bridge this gap by persuading leading authorities in the field to present their thoughts in a straightforward and readable manner.

1 Releasing hormones

H. M. Fraser

This book begins with one of the most exciting areas of current endocrine research, the isolation and characterization of the hypothalamic releasing hormones. From their pioneering work in the 1940s and 1950s Geoffrey Harris and his coworkers postulated that the hypothalamus regulated the secretions of the anterior pituitary gland by liberating substances which were carried to the pituitary via the hypophysial portal blood vessels (see Fig. 1-1 and Book 3, Chapter 1). Termed 'releasing factors', there was thought to be one for each anterior pituitary hormone. It seems surprising today that even as late as the early 1960s their existence was still in doubt. The development of sensitive bioassays for luteinizing hormone (LH), follicle stimulating hormone (FSH) and thyroid stimulating hormone (TSH) enabled people to demonstrate the presence of these factors in the hypothalamus by showing a rise of the relevant pituitary hormone concentration in the blood of rats injected with an appropriate hypothalamic extract. But the precise isolation and subsequent chemical synthesis of these releasing hormones was to prove an enormous task.

ISOLATION OF HYPOTHALAMIC RELEASING HORMONES

The releasing hormone for LH and FSH

Although several laboratories throughout the world were involved, the skill and tenacity of two American groups, one led by Andrew Schally in New Orleans and the other by Roger Guillemin in La Jolla, resulted in the isolation and synthesis of gonadotrophin releasing hormone, work for which they shared the Nobel Prize in 1977. Because the releasing factors

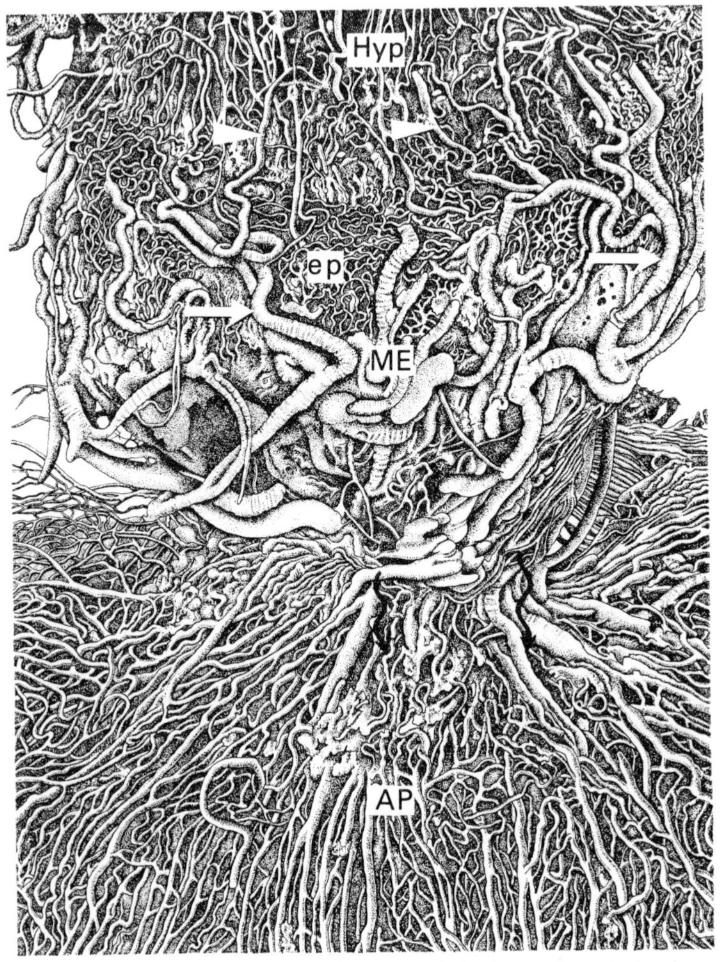

Fig. 1-1. The monkey median eminence and pituitary, frontal view from above. Superior hypophyseal arteries (long white arrows) supply the external plexus (ep) of the median eminence (ME). The median eminence in turn drains to the anterior pituitary (AP) through long portal vessels (wavy arrow). The superior hypophyseal arteries supply not only the median eminence external plexus but also the medial basilar hypothalamus (HYP) through ascending arterioles (white arrowheads). (From R. B. Page *et al.* The neurohypophyseal capillary bed. II. Specializations within median eminence. *Am. J. Anat.* **153**, 33 (1978).)

were present in such small quantities, their isolation required the extraction of hundreds of thousands of pig hypothalami (Schally's group) or sheep hypothalami (Guillemin's group), the mere collection of which required considerable organization.

The hypothalamic substances controlling pituitary gonadotrophin release were the most sought after; initial analysis of the LH-releasing hormone showed it to be a basic peptide. Schally's chemists worked on its structure using 800 μg of purified peptide prepared from 160000 pig hypothalami by acid extraction and various chromatographic steps. The amino acid composition was determined on acid hydrolysates, and it was found to consist of ten amino acids. The molecule was digested with chymotrypsin and thermolysin and the amino acid sequence determined by means of the Edman-dansyl procedure coupled with the selective tritiation method for C-terminal analysis. Schally was the first to propose the structure shown in Fig. 1-2 in 1971; this was confirmed by mass spectral fragmentation, and an identical structure was announced a few months later by Guillemin's group.

The work on the isolation of the LH-releasing factor had

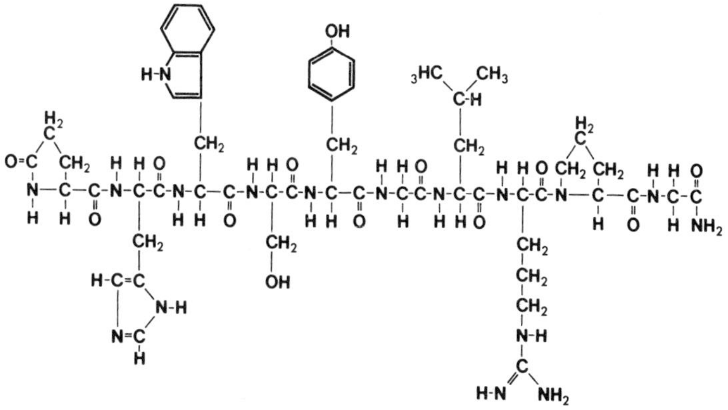

pyroGlu–His–Trp–Ser–Tyr–Gly–Leu–Arg–Pro–Gly-NH_2

Fig. 1-2. Structure of gonadotrophin releasing hormone isolated from ovine and bovine hypothalami. (From R. Guillemin. *Contraception* **5**, 1 (1972).)

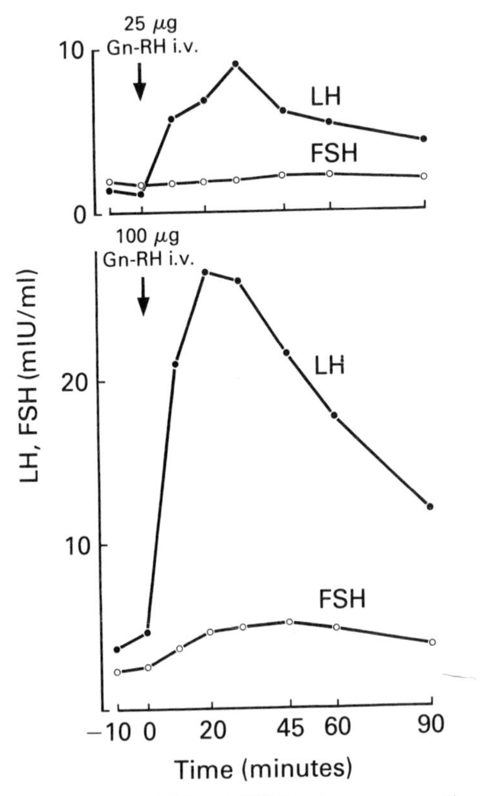

Fig. 1-3. Mean response of LH and FSH to intravenous (i.v.) injection of 25 or 100 μg of Gn-RH in normal men. (From P. Franchimont *et al. Clin. Endocrinol.* **3**, 27 (1974).)

progressed in the belief that a separate releasing factor existed for FSH. However, the FSH-releasing activity of hypothalamic extracts could not be separated from the LH-releasing activity; the one decapeptide released both FSH and LH. This resulted in some confusion over the nomenclature of the decapeptide, so that it was variously known as luteinizing hormone releasing factor (LHRF), luteinizing hormone releasing hormone (LH-RH), LRH, LH/FSH-RH, gonadotrophin releasing hormone (Gn-RH) and, as if that were not enough, a recent international committee on nomenclature have christened it Luliberin. Since most people agree that the term *factor* should be replaced by *hormone* once the structure of a substance has been established, we will refer to this decapeptide as Gn-RH throughout this book since it releases both LH and FSH.

Is there a separate FSH-RH? Arguments have raged over the years but nearly all the chemical evidence is now against this idea. However, under some physiological conditions – e.g. during the onset of puberty in the male and female, and at some stages of the oestrous or menstrual cycle – the LH:FSH ratio may change. In addition, while an injection of Gn-RH leads to a rapid sharp rise in LH in the blood, the release of FSH is usually much less pronounced (Fig. 1-3). In 1971, Schally and his colleagues put forward the view that Gn-RH controlled the release of both LH and FSH and suggested that differences in the LH:FSH ratio might be explained by feedback effects of steroids on the pattern of Gn-RH release and on the response of the pituitary gonadotrophs. While this is still the most plausible hypothesis, we lack direct proof. Yet another factor that may regulate the LH:FSH ratio is 'inhibin'. Obtained from ovarian or testicular extracts, injection of 'inhibin' results in a preferential decline in FSH release. 'Inhibin' has still to be characterized, and even its existence is a matter of some dispute.

Hypothalamic control of prolactin secretion

The classical experiments on the hypothalamic control of the anterior pituitary gland showed that when the link between the hypothalamus and the pituitary was cut, or when the pituitary was transplanted under the kidney capsule, the blood gonadotrophin levels declined, whereas prolactin levels increased. This strongly suggested that prolactin, unlike the gonadotrophins, was normally under the influence of a hypothalamic prolactin inhibitory factor (PIF). Nobody has yet succeeded in isolating a peptide with PIF properties, and it appears to be a biogenic amine, probably dopamine (Fig. 1–4), secreted from nerve terminals in the median eminence into the hypophysial portal blood.

There may also be a prolactin releasing factor (PRF). Thyro-

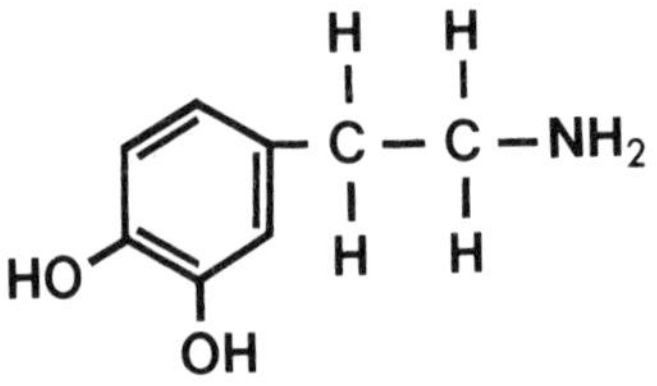

Dopamine

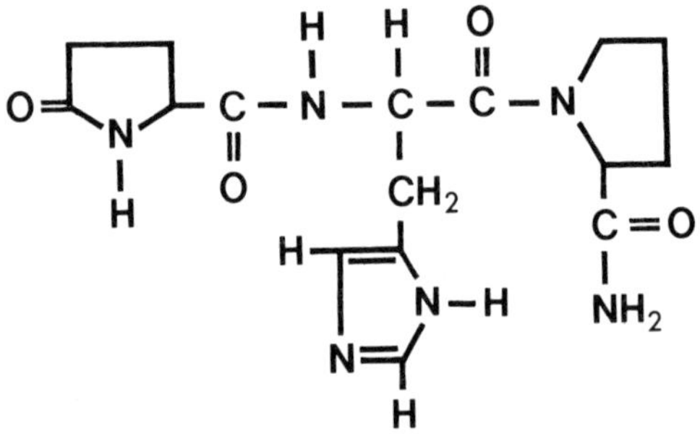

pyroGlu–His–Pro–NH_2 (TRH)

Fig. 1-4. Structure of dopamine and thyrotrophin releasing hormone (TRH). Dopamine, released from the hypothalamus into the hypophysial portal blood, is thought to be the principal prolactin inhibitory factor. As well as releasing TSH, TRH might also have a role in stimulating prolactin release from the pituitary.

trophin releasing hormone (TRH) was the first releasing hormone to be isolated, again by the groups of Schally and Guillemin in 1969; it is a tripeptide, and its structure is shown in Fig. 1-4. As well as causing the release of TSH, it can also induce the release of prolactin *in vivo* and *in vitro*. The significance of TRH in prolactin secretion under physiological conditions is difficult to decide. There have also been reports of another peptide in extracts of porcine hypothalami that releases prolactin, but it has still to be identified.

Physiological effects of Gn-RH

Once the structure of Gn-RH was established, chemists rapidly synthesized it using either classical methods or solid phase synthesis. Thus, the hormone that had taken years to prepare and was available in minute amounts to only a few dedicated workers could now be made available to all. This initiated an enormous amount of research. Not only was this new hormone of great scientific interest, but it also held the key to new methods of fertility control. Since Gn-RH appears to lack species specificity, the one decapeptide was biologically active in man, monkeys, cattle, rabbits, hamsters, rats, mice and birds, as well as in the pig and sheep from which it was first isolated.

Gn-RH is active when administered by a variety of different routes – intravenous, intramuscular, subcutaneous – and is even effective when given by nasal spray, although much larger doses are required. It is ineffective if given orally, as it is rapidly digested before being absorbed into the blood stream. Dosages of only 100 ng/kg are effective by injection, and under many circumstances a response can be induced by even lower doses. However, the duration of the biological effect is very short-lived, being only an hour or so, and the hormone has a half-life of 4–15 minutes in the circulation.

The most obvious physiological action of Gn-RH is its ability to stimulate the pituitary to induce the ovulation of a suitably primed Graafian follicle. Thus there was a hope that it could be used to mimic normal gonadotrophin output in anovulatory women and induce single ovulations. So far this has not proved possible since repeated injections are required; in future, repeated use of the nasal spray, or the use of longer-acting synthetic Gn-RH analogues, may be more successful. Gn-RH could also prove useful for the induction of ovulation in domestic animals. Already, ewes can be made to ovulate during the summer when they are in anoestrus, but unfortunately the resultant corpus luteum is non-functional. From what we now know about the long period of gonadotrophin priming required for follicular development (see Chapter 2), we should not be surprised that a single injection of Gn-RH is not a good way of inducing ovulation.

The induction of infertility by interfering with the action of Gn-RH was first demonstrated by using antibodies to Gn-RH. Conjugating Gn-RH to a carrier protein can make it antigenic; the conjugate is injected into animals as a water-in-oil emulsion with Freund's complete adjuvant. Antibodies to the Gn-RH are produced slowly at first, and then the titre builds up after about 8 weeks. The antibodies cross-react with the animal's own Gn-RH and inhibit it, thereby leading to a reduction in pituitary synthesis and secretion of LH and FSH (Fig. 1-5), with consequent spermatogenic arrest in the male (Fig. 1-6) and ovulatory failure in the female. Such immunization experiments not only confirm the physiological role of Gn-RH, but they also demonstrate one way in which Gn-RH may be manipulated to depress fertility.

ACTIONS OF RELEASING HORMONES

The target cell

The study of the mechanisms of action of the hypothalamic releasing hormones has not progressed as rapidly as the work on other peptide or steroid hormones. The fact that the releasing hormones have been available for a shorter time is obviously one reason for this, but perhaps another problem is that the target organ, the anterior pituitary gland, contains cells that secrete at least six different hormones with the gonadotrophs making up only about 5 per cent of the cell population. It is difficult to obtain purified preparations of pituitary cells secreting only one hormone. The mammotrophic cells secrete prolactin alone, but whether LH and FSH are secreted by a single gonadotroph cell, or by two separate cells, has still to be established.

Not surprisingly, the releasing hormones induce both the synthesis of the appropriate pituitary hormones as well as their release. Addition of Gn-RH to anterior pituitary cells incubated *in vitro* increases the LH and FSH content in the cells and in the medium. Gn-RH action would therefore be expected to induce rapid release of LH and FSH, plus some degree of synthesis. After prolonged infusion of Gn-RH, the pattern of secretion of LH suggests the initial release of hormone from a readily releasable pool followed by release of stored or newly synthesized hormone from a second pool. There is no structural evidence for two different compartments in the cell and the concept of two pools is really a simple way of explaining the hormonal patterns we see. In fact, these pools reflect complex biochemical mechanisms that we do not yet understand.

Most of our knowledge about the synthesis of the pituitary hormones is based on information obtained from other, more easily studied, protein hormones, so that mechanisms that may be peculiar to the anterior pituitary gland are less well understood. We assume that the process involves the standard steps in protein synthesis of transcriptional changes and synthesis of mRNA. Synthesis takes place on the ribosomes of the rough endoplasmic reticulum and the pituitary hormones are then concentrated into secretory granules in the stacked Golgi cisternae. The hormones are released from the cell by a process of exocytosis, in which the membrane enclosing the secretory granule is first fused with the plasma membrane, and the granule is then extruded. The passage of granules within the cell may involve a microtubule–microfilament system.

Attachment to membrane receptor

The first stage in the action of a hypothalamic releasing hormone, like that of any other peptide hormone, is to combine with its specific receptor on the plasma membrane of the target cell. The receptor has two important functions: it recognizes and combines with the hormone, and then transmits a signal to the region of the cell where its biochemical function will be initiated.

Plasma membrane binding in the pituitary has been demon-

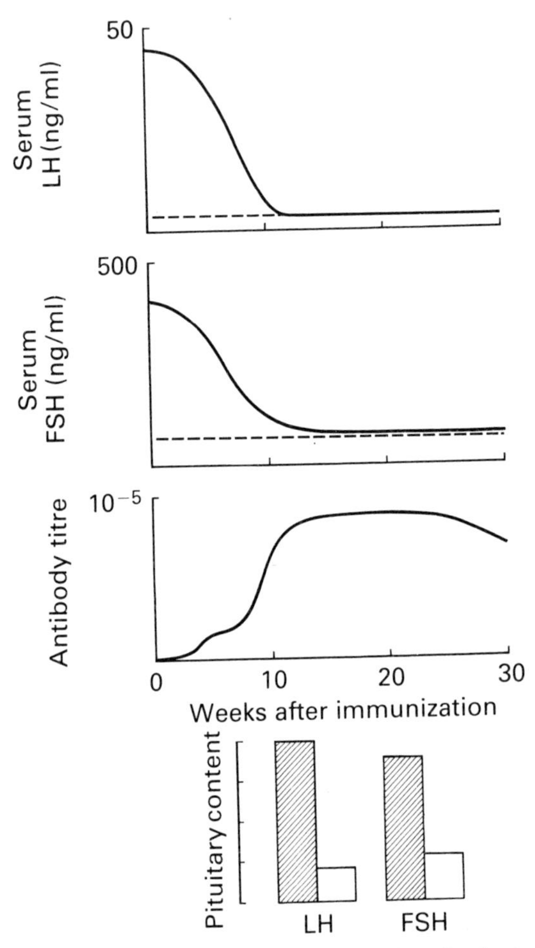

Fig. 1-5. Diagram showing rise in Gn-RH antibody titres and decline in levels of LH and FSH in the blood a few months after immunization against Gn-RH in the male rat. The anterior pituitary glands of the rats have lower levels of LH and FSH (open bars) than controls (hatched bars) indicating decreased synthesis of these hormones. The dashed line indicates sensitivities of the assays. Antibody titres gradually decline after about 20 weeks, and once they are low enough, levels of LH and FSH in the blood will eventually begin to rise unless a booster immunization is given. (Adapted from H. M. Fraser and J. Sandow. *J. Endocrinol.* **74**, 291 (1977).)

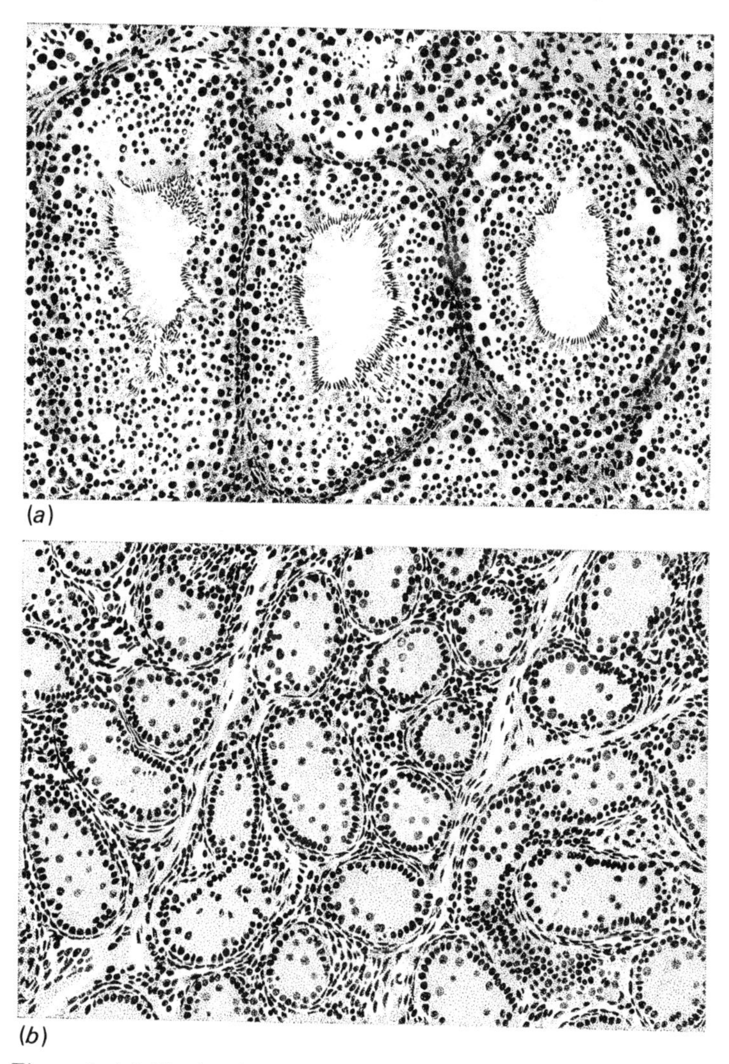

Fig. 1–6. (*a*) Testis of control rabbit showing spermatogenesis progressing normally. (*b*) Testis of rabbit producing antibody to Gn-RH, 4 months after immunization, at the same magnification. Note the small size of the tubules and how spermatogenesis has virtually ceased. (From H. M. Fraser and A. Gunn. *Nature* **244**, 160 (1973).)

strated for TRH and Gn-RH. The membrane fraction is prepared by differential centrifugation of homogenates from anterior pituitary cells; it is then incubated with pure synthetic releasing hormone labelled with a radioactive marker. Incubation takes place between 0 °C and 4 °C to reduce enzymatic breakdown of the hormone, and after a given time the plasma membrane is recovered by centrifugation, and binding assessed by counting the associated radioactivity. Provided the peptide has not been damaged by the radioactive labelling procedure it will bind rapidly (Fig. 1-7), and the binding increases with the

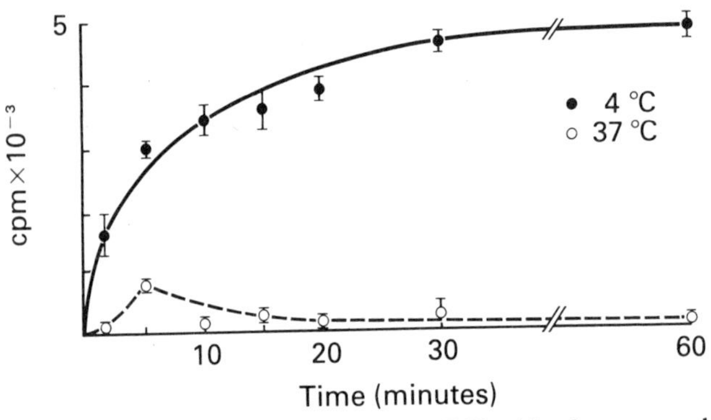

Fig. 1-7. Association of ^{125}I-labelled Gn-RH with plasma membrane fractions of anterior pituitary glands of rats as a function of time and temperature. (From J. Spona. In *Basic Applications and Clinical Uses of Hypothalamic Hormones*, ed. A. L. Charro Salgado *et al*. Amsterdam; Excerpta Medica (1976).)

amount of hormone added until all the receptors are occupied and a plateau is reached. The specificity of binding must be checked by making sure that the bound hormone cannot be displaced by other hormones, which in turn should not be able to bind to the same receptors. Fig. 1-8 shows how the labelled Gn-RH can be displaced with unlabelled Gn-RH whereas other hormones have little or no effect. Finally, the strength or affinity of binding is assessed from a Scatchard plot analysis, in which the ratio of bound to free labelled hormone is plotted against the amount of labelled hormone bound. This has revealed one

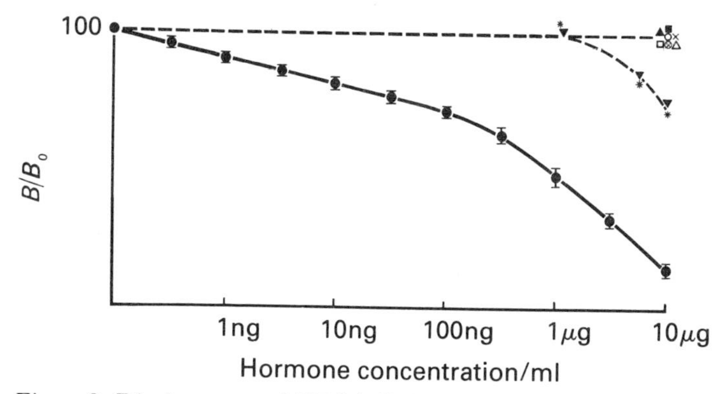

Fig. 1-8. Displacement of ^{125}I-labelled Gn-RH binding to the membrane fraction of rat anterior pituitary gland by Gn-RH. Values are plotted as per cent of the initial binding. ● Gn-RH; □, LH human; ○, FSH human; △, TSH human; ×, ACTH human; ⊗, TRH; ▲, lysine vasopressin; ■, angiotensin II; ▲, gastrin; ✱, arginine vasopressin. (From J. C. Marshall *et al. Clin. Endocrinol.* **5**, 671 (1976).)

binding site for TRH, and two binding sites for Gn-RH, one with low affinity and one with a high affinity. The low affinity site may be concerned with FSH release, although some workers have also found the low affinity receptor in other tissues so that it may be of no particular physiological significance in the pituitary.

Gonadal steroids may possibly regulate gonadotrophin release by changing the availability of Gn-RH receptors in the pituitary. There is some evidence that Gn-RH receptors increase on the afternoon of pro-oestrus in the rat, but this subject has not received much attention in the past. Now that we have available both stimulatory and inhibitory analogues of Gn-RH, this should encourage more work on receptor numbers. Labrie's group in Quebec have shown an increase in TRH binding sites after treating rats with oestrogen, which increases the secretion of prolactin and TSH.

If the hypothalamic releasing hormone receptors work by a similar mechanism to those of other protein hormones, then only a small proportion needs to be occupied for full activation of the adenylate cyclase system (Chapter 2). An increase in the number of binding sites would therefore result in a similar response to a lower amount of releasing hormone. Possibly the releasing hormones can also exert an autoregulatory effect when in high concentrations by being able to decrease the number of available receptors, as has been shown for LH and other polypeptide hormones (see Chapter 2).

Nothing is known about the chemical composition of the Gn-RH receptors, but they are probably similar to those for other protein hormones. The proteins of the plasma membrane are either loosely associated with its periphery (peripheral proteins), or tightly bound to the membrane (integral proteins). Some integral proteins extend into the hydrophobic region of the lipid bilayer, or even span the membrane completely; hormone receptors are probably of this kind. A hydrophilic region of the molecule protrudes from the membrane into the external aqueous phases and combines with the hormone.

Role of cyclic AMP

Once the releasing hormone has bound to its membrane receptor it is thought to activate the enzyme adenylate cyclase which converts ATP to cyclic AMP + PP_i. The cyclic AMP produced presumably stimulates the extrusion of preformed hormone from storage granules across the cell membrane. The releasing hormones themselves are unlikely to act at the genomic level; probably the cyclic AMP acts as an intracellular mediator to stimulate the transcriptional changes leading to synthesis of new hormone.

The mechanism by which the hormone–receptor complex activates adenylate cyclase is still obscure, despite the fact that it is involved in the mechanism of action of many hormones (see Chapter 2). The idea has been put forward that once the hormone–receptor complex is formed, this causes the complex to associate with the enzyme by lateral diffusion through the cell membrane.

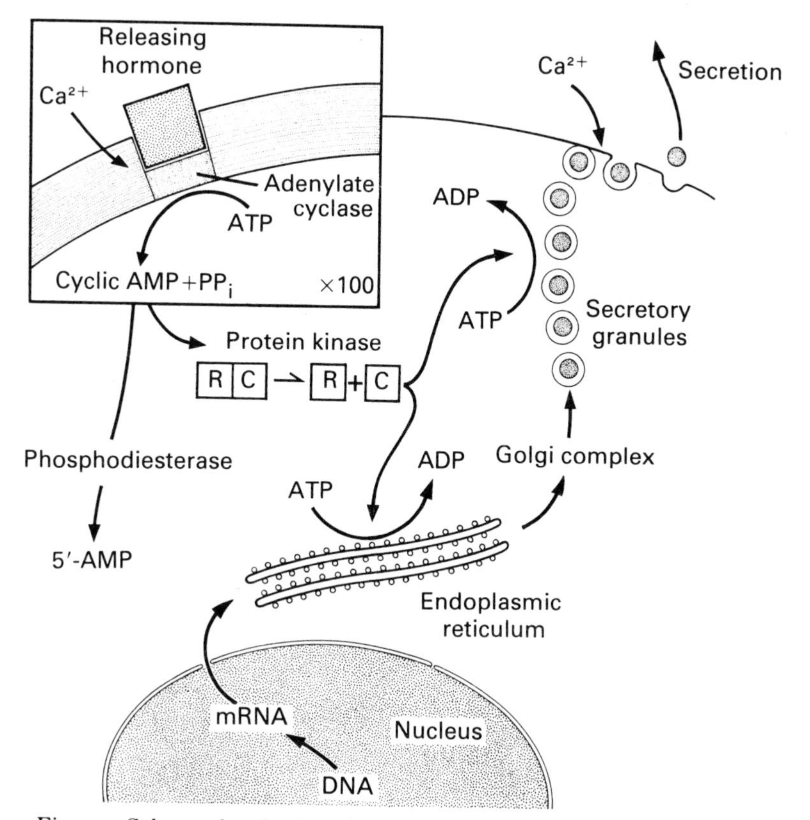

Fig. 1-9. Scheme showing how hypothalamic releasing hormones could act in a cell of the anterior pituitary. The releasing hormone binds to a receptor in the plasma membrane and activates adenylate cyclase thereby stimulating production of cyclic AMP. The cyclic AMP would bind to its receptor and release the activated catalytic subunit (C) of protein kinase from its regulatory subunit (R). By the process of phosphorylation, proteins are activated to control cell function. (Adapted from F. Labrie *et al.* In *Subcellular Mechanisms in Reproductive Neuroendocrinology*. Ed. F. Naftolin *et al.* Amsterdam; Elsevier (1976).)

Many years ago cyclic AMP was shown to increase both the synthesis and release of LH and FSH *in vitro*. Its direct involvement in the action of Gn-RH was suggested by the work of Labrie and his colleagues who found that addition of Gn-RH to male rat hemipituitaries *in vitro* increased levels of cyclic

AMP. Furthermore, synthetic agonists of Gn-RH also increased cyclic AMP in amounts that were related to their ability to release LH and FSH, while antagonists of Gn-RH decreased the formation of cyclic AMP.

Our knowledge of the precise mechanism of action of the hypothalamic releasing hormones is still scanty, and there are many factors whose involvement is still obscure. For example, calcium is required for the action of Gn-RH on cyclic AMP, and although it is also involved in the action of other protein hormones (see Chapter 2), its site of action is unclear. High concentrations of potassium ions stimulate the secretion of hormones from the anterior pituitary gland *in vitro*, suggesting that changes in membrane potential or permeability may be associated with the secretory process. As discussed in Chapter 3, even prostaglandins can induce an LH release, although as we shall see later their action seems to be primarily mediated by increasing the output of Gn-RH.

Fig. 1-9 shows a representation of releasing hormone action as proposed by Labrie and his colleagues.

STRUCTURE–ACTIVITY RELATIONSHIPS OF Gn-RH

Fragments of Gn-RH

Study of the conformation of a hormone in solution and the functional groups most important for its biological action are a first step in determining the way in which it binds to its receptor. Gn-RH is thought to have a conformation something like a horse-shoe, but no one has so far been able to work out exactly which amino acids are responsible for binding to the receptor, and which trigger the biological response. The complete molecule is necessary for biological activity, as removal of any one of the amino acids results in a negligible release of LH and FSH. In addition to studies with fragments of Gn-RH, several hundred analogues have now been synthesized in an effort to produce agonists and antagonists of Gn-RH for the stimulation or inhibition of fertility.

Synthetic agonists of Gn-RH

Gn-RH agonists, analogues that are more potent than Gn-RH in stimulating LH and FSH release, have been developed for the treatment of hypogonadal states and infertility. Although the natural hormone can be used for this purpose, it has to be administered very frequently because of its short biological effect of only a few hours. Agonists have been produced by substituting different amino acids in the Gn-RH molecule and comparing the LH/FSH releasing activity with an equal dose of Gn-RH in various test systems. The best agonists have been derived by modifying either Gly^6, Leu^7 or $Gly\text{-}NH_2{}^{10}$, and the reason for this becomes apparent if you consider the possible sites of enzymatic breakdown of the Gn-RH molecule (Fig. 1-10). By

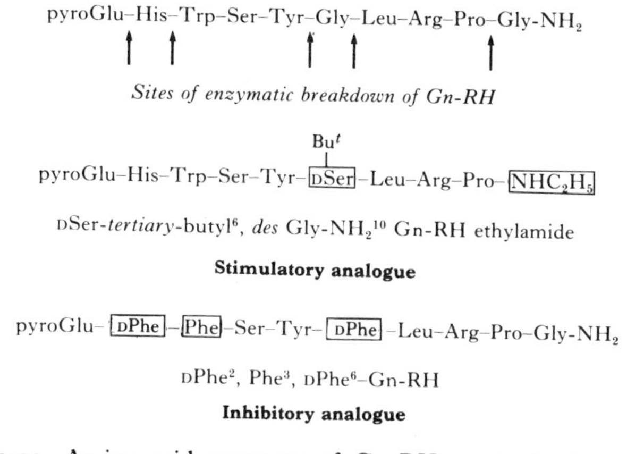

Fig. 1-10. Amino acid sequence of Gn-RH, and stimulatory and inhibitory analogues. Arrows show sites of enzymatic breakdown of Gn-RH. Note amino acid substitutions that produce stimulatory or inhibitory analogues.

substitution with groups that are more resistant to enzymatic attack, not only in the anterior pituitary gland, but also in the kidney, liver and other tissues, one can significantly increase the duration of the biological effect of the hormone. The first really effective substitution was described by Fujino and his colleagues

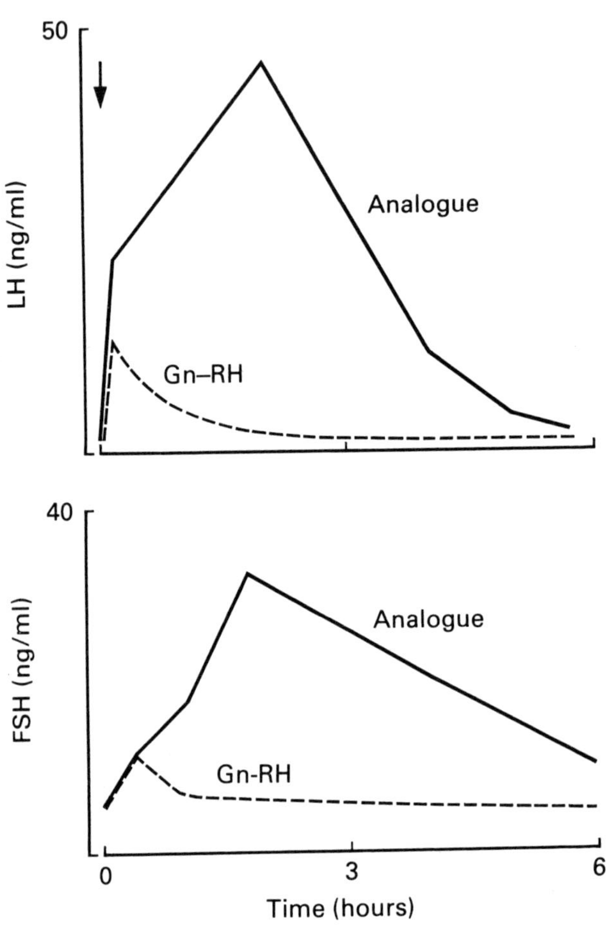

Fig. 1-11. Plasma levels of LH and FSH after an injection of 100 ng Gn-RH or [DSer(Bu^t)^6 *des* Gly^{10}]Gn-RH ethylamide in the rat. (From J. Sandow. In *Basic Applications and Clinical Uses of Hypothalamic Hormones*, Ed. A. L. Charro Salgado *et al.* Amsterdam; Excerpta Medica (1976).)

in Japan, who replaced Gly-NH_2^{10} by an ethylamide group. The LH/FSH releasing activity of this analogue can be enhanced still further if a branched side-chain D-amino acid is substituted for Gly^6. The addition of a protective group such as tertiary butyl on the D-amino acid makes the molecule even more resistant to enzymic attack (Fig. 1-10). Such an agonist will not only produce LH and FSH release in much smaller doses than Gn-RH, but it will also stimulate release over a much longer time period (Fig. 1-11). Agonists currently being used are thirty to one hundred times as active as Gn-RH.

How is this brought about? The biochemical mechanisms are currently being investigated. Although the half life in the blood is known to be the same as that of Gn-RH, these agonists are more resistant to degradation by tissue enzymes so they may persist for longer on the target cell. In addition the agonists have been shown to have a more prolonged pituitary uptake than Gn-RH, and they possibly have a greater affinity for the Gn-RH receptor.

Synthetic antagonists of Gn-RH

The main aim of much of the more recent research in reproductive biology has been to find new and better methods of controlling human fertility. Since Gn-RH release initiates ovulation, an obvious method of preventing ovulation would seem to be by administering a chemical antagonist to Gn-RH. The development of such antagonists followed similar lines to those already outlined for the development of stimulatory analogues, but in this case the end-point was the ability to inhibit the LH release induced by injection of the natural decapeptide. Basically, the antagonist must have a particularly high affinity for the Gn-RH receptors, thereby occupying them for a long time and thus excluding endogenous Gn-RH. At the same time the inherent LH/FSH releasing activity of the antagonist must be as low as possible. The best antagonists have been produced by modifying positions His^2, Trp^3 and Gly^6, especially by replacement with D-amino acids having aromatic side chains (see Fig. 1-10).

We still need to develop more potent antagonists if they are to have any clinical application as potential contraceptives in women. A dose of 200 μg of the existing compounds is still required even to inhibit ovulation in rodents. In the rat and hamster the antagonist can be administered at the optimal time, i.e. during the occurrence of the preovulatory LH surge. Since the precise timing of this surge is not known for women, an antagonist with a very long biological half-life is required, so that Gn-RH can be inhibited for a 2–3-day period to prevent the LH surge, while in contraceptive practice it would have to be taken for several days to prevent follicular development and unacceptably high oestrogen secretion.

This is a particularly exciting and challenging area of research which is becoming a battle-ground between the scientist and the

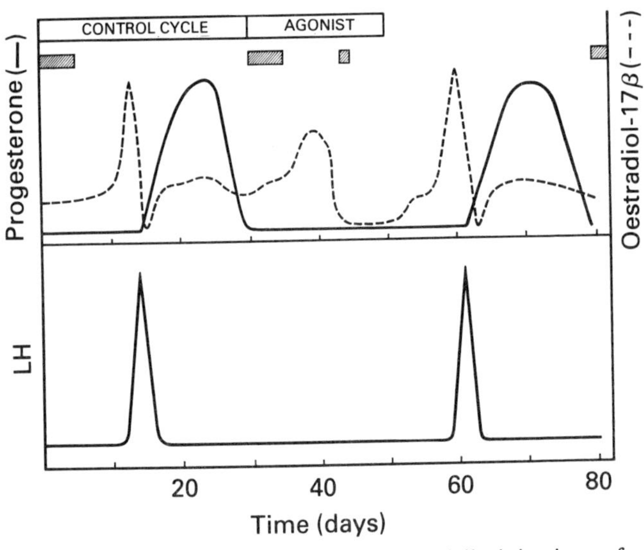

Fig. 1-12. Diagram showing the effect of 20 daily injections of 5 μg Gn-RH agonist beginning on day-1 of the cycle to a stump-tailed macaque (bars indicate menstruation). The basal hormone levels, taken immediately before the daily injection of agonist, show how the levels of oestradiol begin to rise but the LH surge does not occur and the follicle fails to ovulate. When levels of oestrogen fall in the absence of progesterone secretion breakthrough bleeding can occur. Once treatment is stopped the next follicle develops normally. (H. M. Fraser, unpublished.)

pituitary. And the pituitary can do some unpredictable things. For example, when the stimulatory analogues of Gn-RH were given chronically to male rats and dogs in large doses during toxicity trials, the testes involuted and spermatogenesis ceased, while female rats similarly treated either stopped cycling, or aborted. This means that the treatment regime for infertile human patients must be very carefully selected so that the agonist is not administered in too high a dose or too frequently.

This potential obstacle to the clinical use of agonists has been turned to advantage as a possible method of contraception. Sven Nillius in Uppsala has found that ovulation can be inhibited in women by administering the agonist by means of a nasal spray. In Edinburgh we have used macaque monkeys with a menstrual cycle similar to that of women to study the mechanism of the inhibitory effects of the agonist on ovulation, and to check for any side effects of the treatment. When the monkeys are given subcutaneous injections of only 5 μg of agonist each day for 20 days starting on day 1 of the cycle, the LH surge fails to occur and ovulation is prevented, as shown by the low levels of progesterone (Fig. 1–12). Once the treatment is stopped, a new follicle starts secreting oestrogen and a normal ovulatory cycle ensues. The mechanisms involved are complex. At first the agonist stimulates release of supraphysiological amounts of gonadotrophins (Fig. 1-11) and this in turn stimulates oestrogen secretion by the ovary. This happens immediately after each injection (not shown in Fig. 1-12) but soon the pituitary shows a decreased responsiveness to the agonist and to endogenous Gn-RH and the ovary becomes less responsive to the gonadotrophin stimulation. This 'down regulation' phenomenon is a common response by target organs to hyperstimulation by a trophic hormone and is associated with a loss or internalization of receptors (see Chapter 2). We suspect that Gn-RH agonist treatment causes the ovarian follicle to develop abnormally, and this, coupled with decreased pituitary responsiveness to Gn-RH, means that the preovulatory LH surge, and hence ovulation, fails to occur.

The next stage in this work involves giving the agonist to the monkeys continually over several cycles to see if oestrogen can be kept at low levels to make it acceptable in contraception. Alternatively, if the agonists prove luteolytic in women it might be possible to give them only once during the late luteal phase of the cycle to induce menstruation and hence prevent pregnancy.

SYNTHESIS AND RELEASE OF RELEASING HORMONES

The hypothalamus is the centre where steroidal feedback from the gonads, psychogenic factors and environmental stimuli are all integrated to modulate the synthesis and release of Gn-RH. The releasing hormones are thought to be synthesized in cell bodies in the arcuate nucleus and preoptic area and to be passed down axons to nerve terminals situated in the median eminence, where they are stored prior to release into the hypophysial portal blood vessels. Discharge from neurones probably follows the all-or-none rule for conduction of a nerve impulse (Book 3, Chapter 2). Biogenic monoamines such as dopamine are also synthesized in nerve cells in the brain, and function as neurotransmitters stimulating or inhibiting the release of the hypothalamic hormones. Presumably the gonadal steroids affect the synthesis of Gn-RH, while the neurotransmitters affect its rate of release. Gonadal steroids and environmental and psychogenic factors may influence the activity of monoamine neurones converging on the neurones synthesizing Gn-RH.

There are many gaps in our knowledge as the control mechanisms are extremely complex; it all takes place in a small area of the brain, and several releasing hormones share the same amines as neurotransmitters. The separation of one system from another is achieved by the connections between different neurones containing steroid receptors, releasing hormones and neurotransmitters. Significant advances have been made recently in localizing these neurones.

Biosynthesis of releasing hormones

Unfortunately, little is known about the biosynthesis of the releasing hormones, although the biosynthesis of TRH and Gn-RH can be demonstrated during incubations of hypothalamus. They are most likely synthesized in a similar manner to other peptide hormones such as vasopressin. Thus, a larger prohormone is first synthesized on ribosomes, and then broken down by peptidases to the biologically active hormone. Alternatively, it is also possible, particularly for the small tripeptide TRH, that synthesis does not require a mRNA template and is non-ribosomal, assembly being achieved by specific enzymes.

A further mechanism for controlling the amount of Gn-RH available for release may be by altering the activity of hypothalamic peptidases that can inactivate Gn-RH. Castration reduces the activity of these enzymes and this effect can be reversed by appropriate steroid replacement. The physiological role of these peptidases is not clear, but their activity is possibly reduced to increase the stores of Gn-RH.

Sites controlling gonadotrophin release

In the rat, if the connections between the median eminence and the preoptic-anterior hypothalamic area are cut, the animal will fail to produce an ovulatory surge of LH, but tonic secretion of gonadotrophin continues (Book 3, Chapter 2). This suggests that the nervous input from the preoptic–anterior hypothalamic area is responsible for the preovulatory LH surge, while tonic release of gonadotrophin is controlled from the medial basal hypothalamus. In recent years, Ernie Knobil and his colleagues have carried out a fascinating series of experiments in the rhesus monkey: ovulatory and oestrogen-induced surges of LH and FSH were still present even when the cerebral hemispheres and all brain structures anterior and dorsal to the optic chiasma had been surgically removed (Fig. 1-13). The suggestion has therefore been made that the control of the preovulatory LH

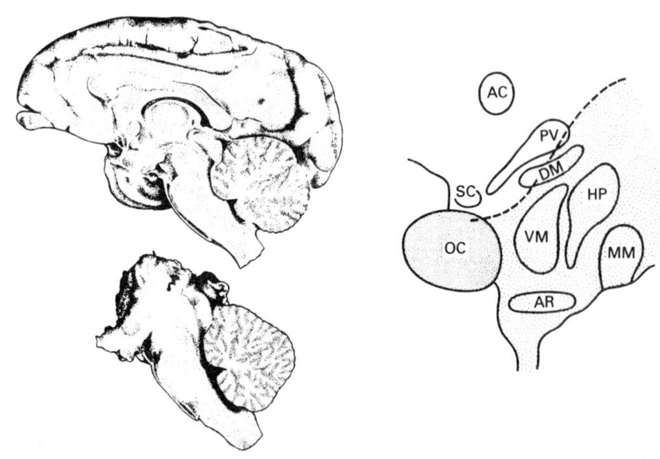

Fig. 1-13. *Left*. Sagittal sections of the brain of a normal female rhesus monkey (above) and one in which the whole of the cerebral hemisphere had been surgically removed, leaving an isolated hypothalamic island (below). Incredibly, this latter animal could still produce an LH and FSH surge in response to oestrogen. *Right*. The diagram shows the areas of the hypothalamus that still remained intact in this operated animal (shaded area). AC, anterior commissure; AR, arcuate nucleus; DM, dorsomedial nucleus; HP, posterior hypothalamic nucleus; MM, mammillary body; OC, optic chiasma; PV, paraventricular nucleus; SC, suprachiasmatic nucleus; VM, ventromedial nucleus. (From D. L. Hess *et al. Endocrinology* **101**, 1264 (1977).)

surge and tonic gonadotrophin secretion originates from the medial basal hypothalamus, or even from the pituitary gland itself. This fundamental difference between the rat and the monkey may reflect the importance of light in controlling the timing of the LH surge in the rodent; evidently the anterior hypothalamic area of the rat receives the light–dark signals that determine the time of ovulation, whereas this diurnal rhythm has no effect on timing ovulation in the monkey.

Localization of Gn-RH and TRH

The ability to produce antibodies to Gn-RH has made it possible to develop two techniques for the precise localization of the areas

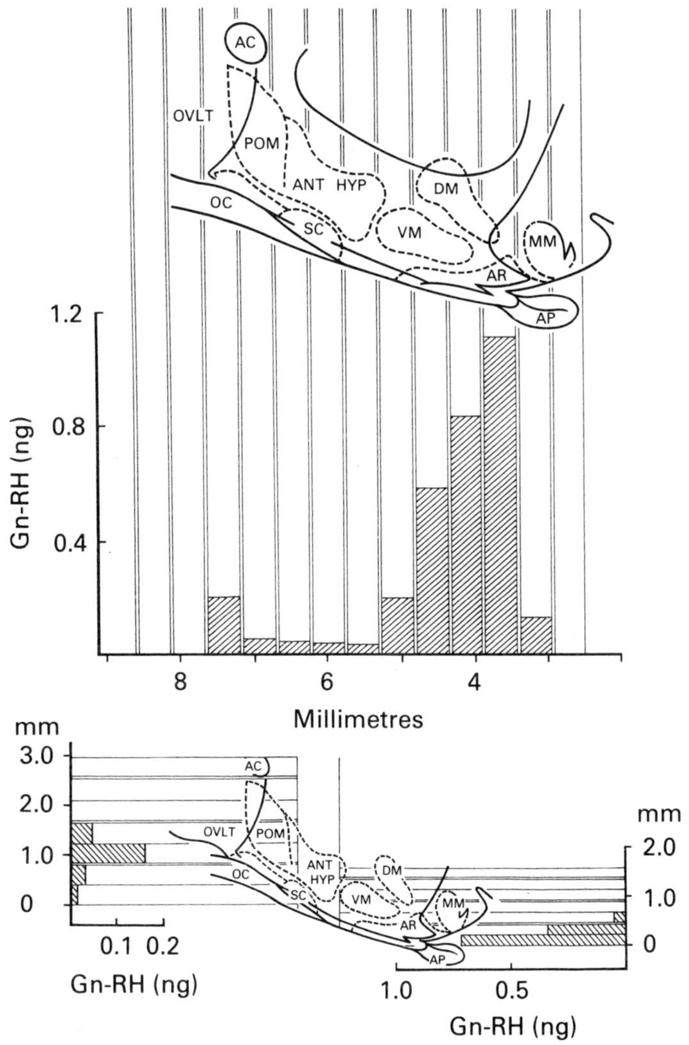

Fig. 1-14. Diagram showing Gn-RH content in frontal and horizontal sections of the rat hypothalamus and preoptic area. AC, anterior commissure; ANT HYP, anterior hypothalamic area; AP, anterior pituitary gland; AR, arcuate nucleus; DM, dorsomedial nucleus; MM, mammillary nucleus; OC, optic chiasma; OVLT, organum vasculosum of the lamina terminalis; POM, preoptic medial nucleus; SC, suprachiasmatic nucleus; VM, ventromedial nucleus. (From J. E. Wheaton *et al. Endocrinology* **97**, 30 (1975).)

of the brain in which Gn-RH is found, and the cellular structures that contain the hormone. These techniques are radioimmunoassay and immunohistochemistry; radioimmunoassay is capable of measuring down to a few picograms (10^{-12} g) of Gn-RH. Serial sections through the brain are cut in the frontal, sagittal and horizontal planes and Gn-RH activity assayed in each section. By projecting the data onto a three-dimensional map, the areas containing the LH-RH can be localized. Fig. 1-14 shows that most of the Gn-RH is stored in the median eminence. But it is also important to consider those areas in which only small amounts of Gn-RH are found, because these could be the sites of synthesis; note that some Gn-RH was found in the medial basal preoptic area, in association with the organum vasculosum of the lamina terminalis. Low levels were also detected between the preoptic area and the arcuate nucleus–median eminence region.

The neurones in which releasing hormones are produced, transported and stored can be examined under the microscope by immunohistochemical techniques. These involve incubating sections of brain with specific rabbit antibody to Gn-RH, which will combine with the hormone. A second antibody raised against the rabbit antiserum is tagged with a fluorescent label and in turn added to the incubation, where it will combine with the releasing hormone–antibody complex. The fluorescence emitted by the complex in ultraviolet light can be viewed under the microscope to show the precise cellular localization of the releasing hormone. This immunofluorescent technique has attracted a tremendous amount of interest. One would expect to find Gn-RH in cell bodies in which it is being synthesized prior to being transported down the axons of the neurones. The cell bodies have been difficult to localize, possible because Gn-RH is present in low concentrations or perhaps some of it exists as a larger prohormone which does not cross react with antibodies to Gn-RH. Those workers who have localized the cell bodies have found them mostly in the preoptic area with a few in the medial basal hypothalamus in the rodent, while in monkeys and in man they appear to be present in equal numbers in the preoptic area and medial basal hypothalamus. Perhaps this difference explains why isolation of the medial basal hypothalamus abolishes the LH surge in the rat but not in the monkey. Certainly, Gn-RH can be demonstrated in the median eminence where it is stored in the axonal terminals prior to release.

Another possible route for the transportation of Gn-RH from cells in the preoptic area into the hypophysial portal blood is via the cerebrospinal fluid (CSF) and the tanycytes. Tanycytes are ependymal cells that line the base of the third ventricle and send basilar processes down to the capillary plexus of the median eminence. It has been postulated that the terminals from some Gn-RH-producing cells end in the third ventricle and release Gn-RH into the CSF. It may then be taken up by the tanycytes and transported to the hypophysial portal blood system. Gn-RH can indeed cause the release of LH when injected into the third ventricle, but whether this route is of any physiological significance is doubtful. For example, most investigators have found little or no Gn-RH in the CSF. An ingenious way of administering Gn-RH is by nasal spray, and the suggestion has even been made that this route is effective because the Gn-RH is absorbed from the nasal mucosa and transported somehow to the CSF, whence it is taken up by the tanycytes and portal blood. However, some must also reach the pituitary by the systemic circulation, and the existence of the CSF route remains unproven.

Gn-RH has also been localized in nerve terminals in areas of the brain where it could not possibly be secreted into the portal vessels. This Gn-RH is thought to act as a neurotransmitter having such functions as influencing sexual behaviour. The fact that TRH, which is present in high concentrations in the median eminence, is also distributed in smaller quantities in several areas of the brain and spinal cord highlights an as yet little known neurotransmitter role for the hypothalamic releasing hormones in addition to their function of controlling release of the hormones of the anterior pituitary gland.

Localization of gonadal steroids in the brain

Receptor sites for oestrogens and androgens have been mapped in the brain of male and female rats but the evidence for progesterone receptors is still equivocal. We may reasonably assume that the mechanism of action of the gonadal steroids in the brain is similar to their actions in other target tissues (see Chapter 5): after binding to a cytoplasmic receptor and being transported to the nucleus, they would alter genomic function so as to influence RNA synthesis and subsequently protein formation.

Presumably the steroids have an effect on the neurones synthesizing Gn-RH and on the activity of neurones containing biogenic amines that act to stimulate or inhibit release of Gn-RH. One might expect to find some steroid receptors in those cell bodies that synthesize Gn-RH and some in neurones synthesizing the biogenic amines. Preliminary evidence suggests this may be the case for Gn-RH but not for the biogenic amines. Some dopamine neurones in the arcuate nucleus are also targets for oestradiol but otherwise hormone uptake sites in the brain overlap little or not at all with amine neurone cell groups, which are found predominantly in the medulla, pons and midbrain. Oestradiol is accumulated largely in neurones in parts of the hypothalamus, septal nuclei and cortico-medial amygdala and these areas are devoid of monoamine neurone cell bodies but rich in terminals. The nature of the link between post-synaptic sites that accumulate sex hormones and their aminergic input remains an important neuroendocrine mechanism to be defined.

Biogenic monoamines

We have known for many years that the biogenic monoamines can affect the secretion of hormones from the anterior pituitary gland. In the brain they are synthesized by a non-ribosomal enzymic mechanism within cell bodies of aminergic neurones. They act as neurotransmitters, and for LH and FSH their action

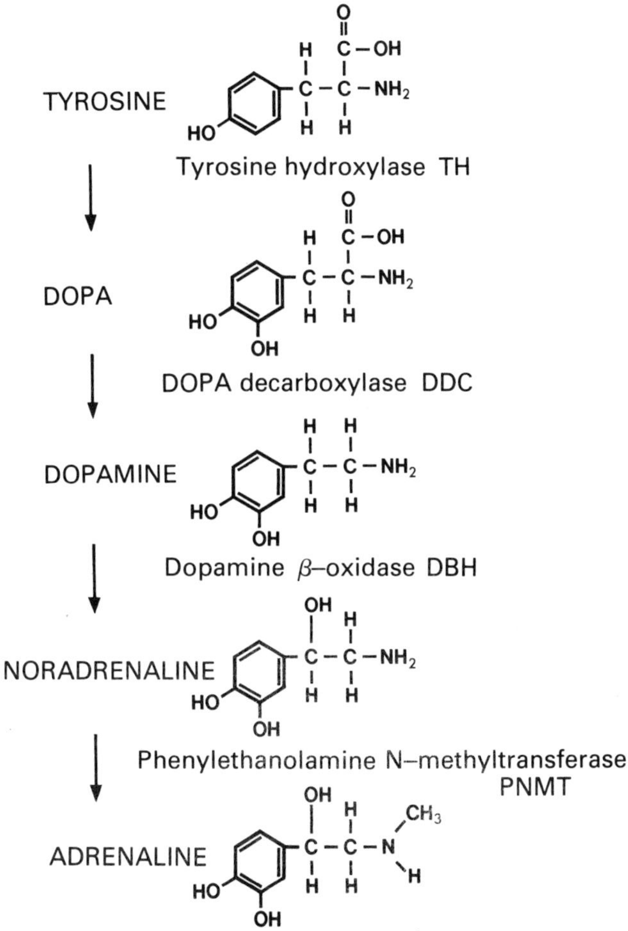

TYROSINE

Tyrosine hydroxylase TH

DOPA

DOPA decarboxylase DDC

DOPAMINE

Dopamine β–oxidase DBH

NORADRENALINE

Phenylethanolamine N–methyltransferase PNMT

ADRENALINE

Fig. 1-15. Pathway of catecholamine biosynthesis from tyrosine. The enzymes controlling each step can be localized by immunohistochemistry. DOPA, dihydroxyphenylalanine. (From T. Hokfelt *et al. Ann. N.Y. Acad. Sci.* **254**, 407 (1975).)

seems to be mediated through influence on the release of Gn-RH, while their effect on prolactin secretion is principally by acting directly on the anterior pituitary gland.

The catecholamines – dopamine, adrenaline and noradrenaline (Fig. 1-15) – and the indolamine, 5-hydroxytryptamine

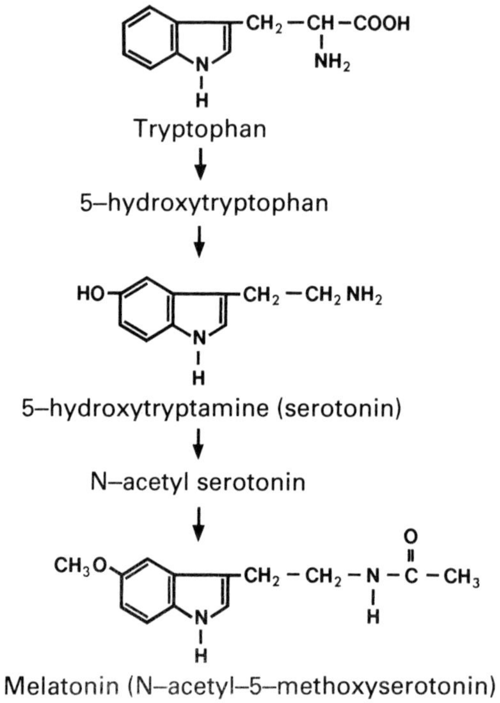

$CH_2-CH-COOH$

NH_2

N

H

Tryptophan

↓

5–hydroxytryptophan

↓

HO $CH_2-CH_2NH_2$

N

H

5–hydroxytryptamine (serotonin)

↓

N–acetyl serotonin

↓

CH_3O $CH_2-CH_2-N-C(=O)-CH_3$ (N–H)

N

H

Melatonin (N–acetyl–5–methoxyserotonin)

Fig. 1-16. Pathway of indolamine biosynthesis from tryptophan.

(serotonin) (Fig. 1-16) seem to be the most important. Their specific effects on the secretion of the anterior pituitary hormones have been studied by a variety of techniques. The effect of the amines themselves can be observed if administered intraventricularly. Alternatively, their precursors can be injected systemically or drugs can be administered to alter the central levels of the transmitters or influence release or uptake processes or receptor activity.

Interpretation of results from different workers is complicated because of the variety of methods used, employment of different dosages and times of exposure and the assessment of response with different endocrine parameters. Both stimulatory and inhibitory roles in gonadotrophin secretion have been claimed

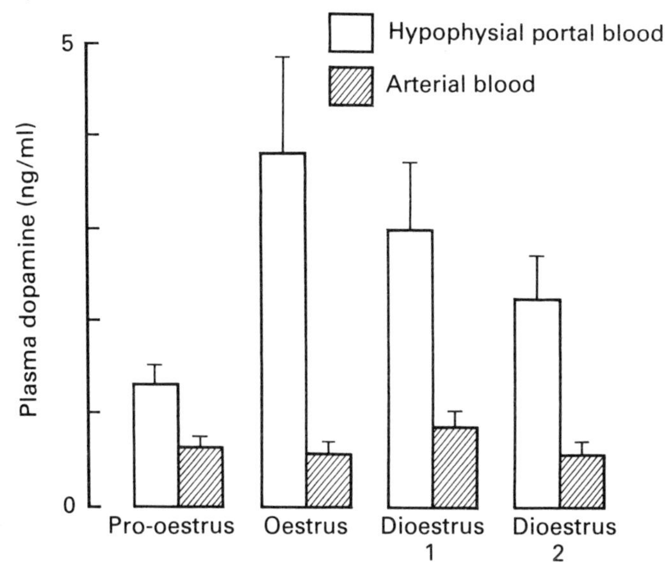

Fig. 1-17. Dopamine concentrations (mean + standard error) in hypophysial portal blood and arterial plasma during the rat oestrous cycle. Note that dopamine was lowest in hypophyseal portal blood at pro-oestrus, when levels of LH and prolactin in the blood are highest. Noradrenaline and adrenaline were undetectable in portal blood as well as arterial plasma. (From N. Ben-Jonathan *et al. Endocrinology*. **100**, 452 (1977).)

for dopamine and 5-hydroxytryptamine, but there is almost universal agreement that noradrenaline stimulates Gn-RH release. When infused intraventricularly in rats it increases Gn-RH concentrations in the hypophysial portal blood, leading to a rise in LH and induction of ovulation.

Although dopamine may be synonymous with PIF (prolactin inhibitory factor) one cannot yet exclude the alternative possibility that it may act by stimulating the secretion of an as yet undefined hypothalamic PIF. During the oestrous cycle of the rat dopamine concentrations in the hypophysial portal blood are lowest at pro-oestrus when prolactin and LH in the blood are elevated (Fig. 1-17). An inhibitory effect on prolactin secretion would appear to be the most dominant physiological control, but some workers have showed that 5-hydroxytryptamine may stimulate prolactin secretion.

The biogenic amines are thought to exert their influence on gonadotrophin secretion by altering the release of stored releasing hormone from nerve terminals. But how can one amine, which has the ability to alter release of several pituitary hormones, be used specifically to regulate the release of only one hormone? We must examine the structural organization of the amine-secreting neurones to see if they are connected to the neurones that secrete the releasing hormones. Techniques for the immunofluorescent localization of the amines and releasing hormones in neurones have been developed by Fuxe and Hokfelt in Stockholm. Since it has not been possible to obtain specific antibodies for each of the amines, antibodies have been raised to the four enzymes that convert tyrosine ⟶ dopa ⟶ dopamine ⟶ noradrenaline ⟶ adrenaline (Fig. 1-15). Thus adrenaline-secreting neurones should contain all four enzymes, whereas noradrenaline neurones should lack phenylethanolamine N-methyltransferase (PNMT), and dopamine neurones should lack both PNMT and dopamine β-oxidase (DBH) (Fig. 1-15). In this way, the various neurones can be differentiated from one another in serial sections. The technique has also been used to study the rate of amine turnover in different reproductive states. A close relationship exists in the lateral external layer of the median eminence between dopamine and Gn-RH-containing nerve terminals (Fig. 1–18), supporting Fuxe and Hokfelt's view that dopamine plays an important role in control of Gn-RH release. The location of noradrenaline neurones suggests they may be involved in the synthesis of Gn-RH. Dopamine turnover decreases and noradrenaline turnover increases in rats when blood LH levels are high, and Fuxe and Hokfelt postulate that oestradiol may inhibit Gn-RH secretion by increasing dopamine turnover in the lateral external layer of the median eminence and reducing noradrenaline/adrenaline turnover. The positive feedback action of oestrogen which induces an LH discharge may involve an increase in noradrenaline turnover and a decrease in dopamine turnover.

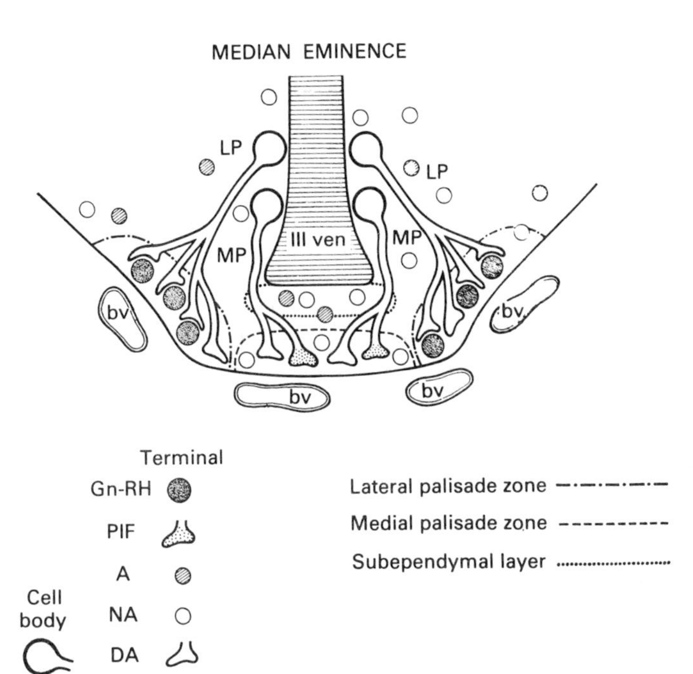

Fig. 1-18. Schematic diagram of the distribution of dopamine (DA), noradrenaline (NA), adrenaline (A) and Gn-RH in the rat median eminence. The lateral DA pathways may inhibit Gn-RH secretion. Some of the medial DA terminals may be concerned with prolactin secretion, the DA acting as a PIF, LP, lateral pathway; MP, medial pathway; bv, blood vessels of primary capillary network; III.ven, third ventricle. (From K. Fuxe *et al.* In *Subcellular Mechanisms in Reproductive Neuroendocrinology*, Ed. F. Naftolin *et al.* Amsterdam; Elsevier (1976).)

Role of prostaglandins

Injection of prostaglandins of the E type (PGE) will induce the release of LH from the anterior pituitary gland. They can act at both the level of the anterior hypothalamus and the pituitary, although most recent evidence suggests that the hypothalamus is the main site of action since administration of PGE increases the level of Gn-RH in the hypophysial portal blood, and prior administration of antibody to Gn-RH will abolish this effect (Chapter 3). PGE is thought to act directly on the Gn-RH neurones independently of the biogenic amines, but its physiological significance is unknown.

Role of the pineal gland

Another organ in the brain that exerts an influence over the output of Gn-RH is the pineal gland. The pineal exerts an effect on gonadotrophin secretion presumably through its production of indoles and peptides of which melatonin may be the most important. These substances are thought to act by inhibiting the output of Gn-RH, and they may pass from the pineal to the hypothalamus via the systemic circulation. The pineal responds to a decrease in day length by increasing its output of melatonin, and melatonin synthesis requires the presence of the neurotransmitter noradrenaline; however, we do not know whether melatonin itself, or some other pineal product, acts to inhibit Gn-RH synthesis or action. The influence of the pineal probably differs considerably between species. Its main role is in the regulation of circannual rhythms so it is particularly important in seasonal breeders. If the sympathetic nerve supply to the pineal gland is cut, the gland regresses and seasonal breeding no longer occurs; Gerald Lincoln in Edinburgh has described this procedure as a way of producing 'a ram for all seasons'.

CONTROL OF PITUITARY GONADOTROPHIN RELEASE

The most powerful factors controlling gonadotrophin release from the pituitary are the feedback effects of the principal gonadal steroids, oestradiol-17β and ovarian androgens in the female, and testosterone and testicular oestrogens in the male. Thus, castration causes a marked increase in gonadotrophin secretion in both sexes, while injection of gonadal steroids will reduce gonadotrophin output (Book 3, Chapter 3).

In the intact male and female, LH is released episodically at irregular intervals (Figs. 1-19 and 1-20). This stimulates the secretion of gonadal steroids which in turn have a negative

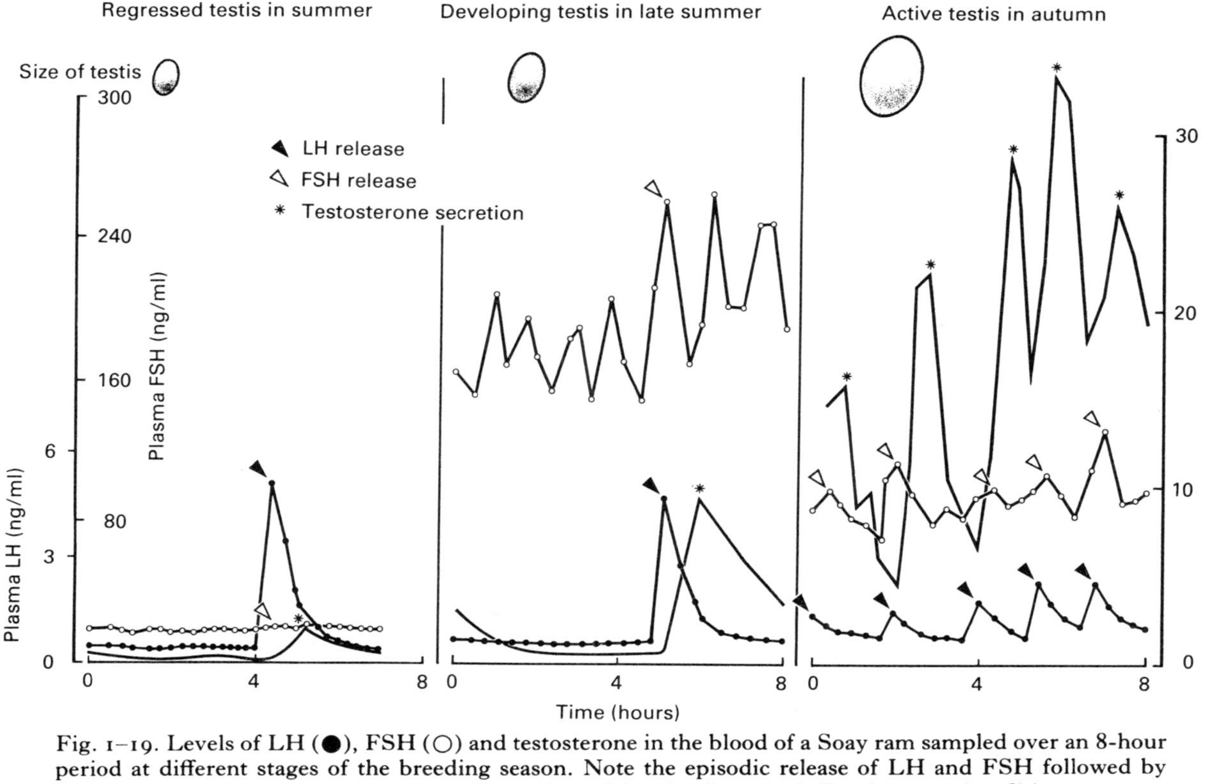

Fig. 1–19. Levels of LH (●), FSH (○) and testosterone in the blood of a Soay ram sampled over an 8-hour period at different stages of the breeding season. Note the episodic release of LH and FSH followed by secretion of testosterone by the testes. (From G. A. Lincoln. *J. Reprod. Fertil.* **53**, 31 (1978).)

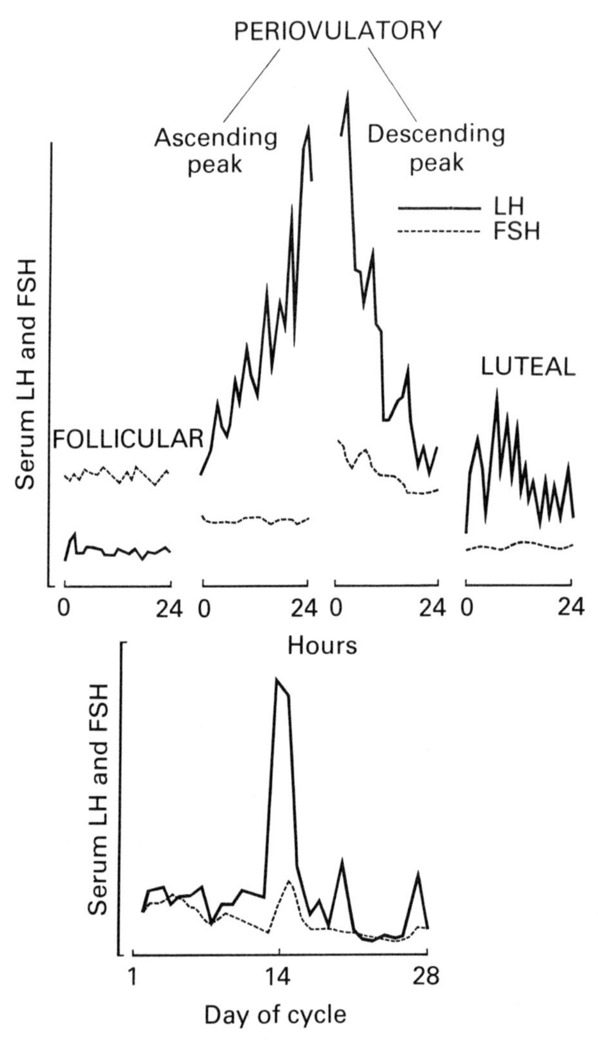

Fig. 1-20. Pattern of secretion of LH and FSH obtained from daily blood samples throughout the menstrual cycle (below) and those obtained from hourly samples at different stages of the cycle (above) in women. (Adapted from A. R. Midgley and P. B. Jaffe. *J. Clin. Endocr. Metab.* **33**, 962 (1971).)

feedback effect. In the normal male, gonadotrophin secretion follows a fairly constant pattern and the relationships between hormones are difficult to disentangle. Exceptions are the seasonal breeders whose reproductive capacities are affected by season and governed by daylength so that they show extremely large seasonal changes in the output of gonadotrophins and in testicular size and function.

When we look at the sequence of hormonal changes in the blood during the reproductive cycles of the female (Book 3, Chapter 3) we find a combination of negative and positive feedback effects, since high concentrations of oestradiol can actually provoke an LH discharge.

The pattern of FSH secretion in both the male and female is somewhat different from that of LH. This can often be attributed to the longer half-life of FSH, but there must also be a mechanism at work that allows a differential pituitary response to the common releasing hormone, Gn-RH. Furthermore, an increased gonadotrophin secretion is brought about not only by altering the secretion of Gn-RH but also by altering the Gn-RH responsiveness of the pituitary to a standard amount of Gn-RH. Most of these changes are themselves mediated by the gonadal steroids, but determining whether the hypothalamus or the pituitary is the more important site of feedback action is proving very difficult.

Hormonal patterns in reproductive cycles

In the female the most difficult thing to explain during the normal cycle is the coexistence of both positive and negative feedback mechanisms. Thus rising levels of ovarian oestrogen will initially depress gonadotrophin secretion by negative feedback and then trigger the preovulatory LH surge by positive feedback. Oestradiol-17β is the most important ovarian hormone in bringing about positive feedback as was demonstrated by Ferin and his colleagues in New York who injected rats and monkeys with antibodies to oestradiol-17β, thereby abolishing the LH surge and inhibiting ovulation.

Early studies to find out how rising levels of oestradiol-17β could produce this switch from negative to positive feedback involved administering different doses of oestradiol during the follicular phase of the cycle to see if an LH surge could be induced. Not surprisingly, injection of oestradiol-17β usually induces negative feedback and stops ovulation. A definitive series of experiments was carried out by Fred Karsh and his colleagues who treated rhesus monkeys with carefully controlled doses of oestradiol-17β during the early follicular phase of the cycle to see which would induce an LH surge comparable with that seen just prior to ovulation. The dose of steroid was gradually increased by means of subcutaneous implants of oestradiol-17β in silastic capsules, varying both the dose and exposure time. Initially a negative feedback effect was observed, but this was followed by a positive feedback effect when the concentration of oestradiol was maintained at 200–400 pg/ml for 36 hours (Fig. 1-21). The results showed that the induction of the LH surge is dependent not only on achieving a certain threshold concentration of oestradiol-17β, but also maintaining it for a critical period of time.

These studies demonstrated the importance of the ovaries in the timing of the LH surge, in contrast to earlier investigations which suggested that the central nervous system possessed some sort of a 'clock' that initiated the LH surge (Book 3, Chapter 2). The clock concept was derived from work done in the rat and hamster in which the LH surge always occurs at a precise time in the afternoon of pro-oestrus and can be altered by changing the lighting regime to which the animals are exposed. Even when these rodents are ovariectomized, oestradiol can still only induce positive feedback in the afternoon. Most people now accept that in these species the oestradiol-17β from the ovaries determines the day of the LH surge, while the actual time of its onset is governed by the diurnal rhythm. Sheep and monkeys show no evidence of diurnal rhythmicity in the timing of the LH surge,

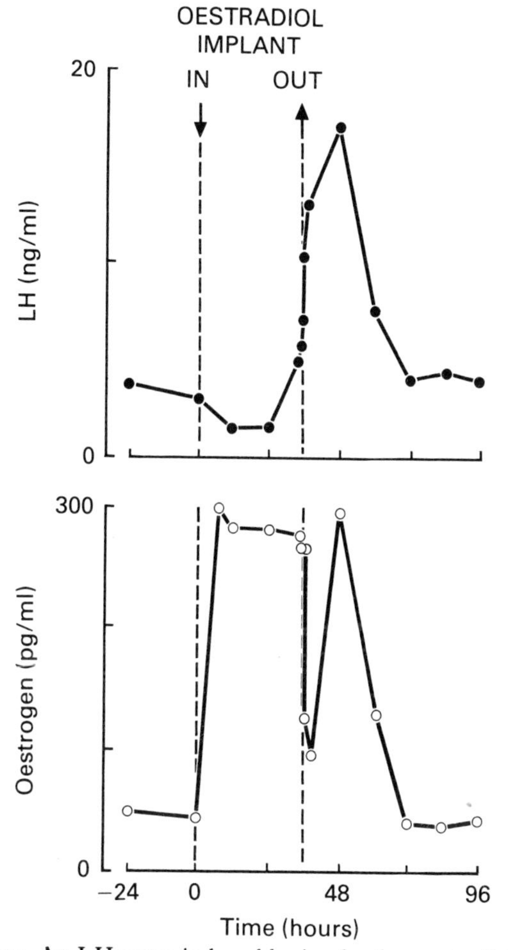

Fig. 1-21. An LH surge induced by implanting oestradiol on day 3 of the menstrual cycle of a rhesus monkey to produce a concentration of 200–400 pg/ml in the blood for 36 hours. The peak of oestrogen associated with the LH surge is the ovarian response to stimulation by the LH. (From F. J. Karsch *et al. Endocrinology* **92** 1740 (1971).)

so in higher species the ovary determines the time of onset of the surge.

The progestins secreted by the ovary are also important in feedback control of gonadotrophins. During the luteal phase of the cycle large quantities of progesterone are secreted by the corpus luteum, and in some species oestradiol-17β is also produced (Book 3, Chapter 3). These steroids exert a negative feedback effect on the release of the pituitary gonadotrophins. Progesterone can also synergize with oestradiol-17β to bring about positive feedback in certain species. In the chicken, progesterone is in fact the principal steroid responsible for the LH surge. In the rat a rise in progesterone of adrenal origin in the afternoon of pro-oestrus facilitates the oestrogen-induced LH surge. In women and rhesus monkeys there is a small rise in progesterone secretion at the time of the pre-ovulatory LH surge and this may play a minor role in augmenting the LH surge.

In the male, the principal control by the testes is generally considered to be via the negative feedback action of the metabolites of testosterone (see Chapter 4) at the level of the hypothalamus where secretion of Gn-RH is reduced. There are indications that 'inhibin', a large molecular weight protein made by the Sertoli cells, acts on the pituitary to reduce FSH secretion selectively but its significance has still to be established.

Gerald Lincoln has studied the Soay ram, a seasonally breeding primitive breed of sheep from the bleak Atlantic island of St Kilda, as a model for the study of the hormonal control of male reproduction. The Soay ram shows large and predictable hormonal changes throughout the year (Fig. 1-19). During long days when the testes are regressed, basal secretion of LH and FSH is very low and the number of pulsatile discharges of LH are few; the testes respond minimally to each LH pulse (Fig. 1-19). As daylength decreases the frequency of LH and FSH pulses and the basal secretion increases, as does the secretion of testosterone by the testes. The decreasing daylight probably causes a decreased hypothalamic sensitivity to the negative feedback effect of testosterone, so that the release of Gn-RH increases. This mechanism is influenced by the pineal gland, particularly in seasonal breeders, and this annual recrudescence of testicular activity is rather similar to the hormonal changes at the onset of puberty. In the sexually active ram the frequency of LH and FSH pulses increases greatly but the amplitude is reduced. At this time the testes show a maximal steroidogenic response to each pulse of LH.

Levels of Gn-RH

Gn-RH is an example of a very potent 'local' hormone that is produced in minute amounts and transported in a very small volume of blood from the hypothalamus to the pituitary before even entering the systemic circulation, where its concentration will become infinitesimal. Even to detect the picogram quantities in portal blood has required the development of a very sensitive radioimmunoassay, and peripheral concentrations still defy detection. Unfortunately, sampling of hypophysial portal blood necessitates major surgery. The rat is the most convenient experimental animal for this purpose but until recently the anaesthetics used in the operation themselves inhibited the LH surge presumably by reducing the output of Gn-RH.

George Fink and his colleagues in Oxford measured Gn-RH in pituitary stalk blood of anaesthetized rats at different stages of the oestrous cycle, and failed to observe any rise in Gn-RH prior to the LH surge. They tentatively suggested that the preovulatory LH surge might be ascribed to the increased responsiveness of the pituitary to Gn-RH on the afternoon of pro-oestrus, rather than to any alteration in Gn-RH secretion. Then in 1976 a new anaesthetic, Althesin, became available; it had been developed for clinical use and was found to have a minimal effect on forebrain function. Using rats anaesthetized with Althesin, Fink was able to demonstrate an unequivocal increase in Gn-RH concentrations prior to the preovulatory surge of LH (Fig. 1-22). There was also a second peak of Gn-RH just after the LH surge, which might be responsible for maintaining the elevated levels of FSH found in rats at this time.

Clearly, there will be difficulties in collecting portal blood

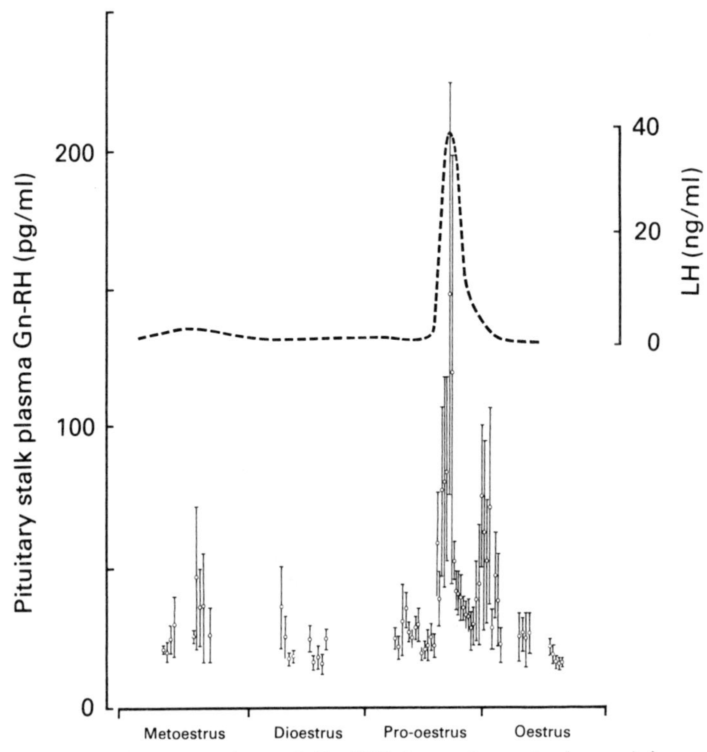

Fig. 1-22. Concentrations of Gn-RH (mean ± standard error) in 30-minute collections of pituitary stalk blood at different stages in the oestrous cycle of the rat. The mean concentrations of LH in the blood are shown for comparison. (From D. K. Sarkar *et al. Nature* **264**, 461 (1976).)

from higher species. The collection can only be maintained for an hour or so in any one animal so that the time of the cycle at which the animal is studied must be known precisely. Fortunately, anaesthetics do not interfere with the hypothalamic function of the rhesus monkey, and recently, elevated levels of Gn-RH have indeed been demonstrated in the hypophysial blood of monkeys at the time of the LH surge.

In species such as sheep and man in which repeated peripheral blood samples can be taken at short time intervals, the LH surge can be seen to be the summation of a series of increasingly rapid pulsatile discharges of LH from the pituitary, but determining whether each of these is in response to a pulse of Gn-RH release will be difficult.

Feedback of gonadal steroids on the pituitary

Receptors for the gonadal steroids are present in the anterior pituitary gland, and studies with pituitaries *in vitro* strongly suggest that steroids have a direct action on the cells to modulate their responsiveness to Gn-RH. Thus, when gonadal steroids are added to pituitaries incubated *in vitro*, both the basal release of LH and FSH, and the LH and FSH released in response to a standard dose of Gn-RH, are reduced, thus suggesting a steroidal negative feedback operating at the level of the pituitary. Demonstrating a positive feedback effect of oestradiol-17β on the pituitary is more difficult since the effect takes a long time to develop, and pituitary tissue will not survive in-vitro incubations for long. Drouin and his colleagues in Labrie's laboratory overcame this by using pituitary cell cultures, and showed that treatment with oestradiol-17β increased their sensitivity to Gn-RH after a lag period of 10 hours, the effect being maximal at 24 hours.

Clearly, we have great difficulty in determining whether the hypothalamus or the pituitary is the more important site of feedback, but these in-vitro studies have established that the pituitary is certainly involved. Eventually, one hopes that feedback effects will explain the changing ratios of LH and FSH secretion as well as variations in the secretion of Gn-RH. But the problem with in-vitro studies is that there are always difficulties in distinguishing between physiological and pharmacological effects.

Changes in pituitary responsiveness to Gn-RH

Studies of the responsiveness of a female animal to a standard injection of Gn-RH at different stages of the reproductive cycle reveal that there are important changes which contribute to the pre-ovulatory LH surge. An increased release of LH and FSH is observed during the late follicular phase in women and other primates (Fig. 1-23) and on the afternoon of pro-oestrus in rats

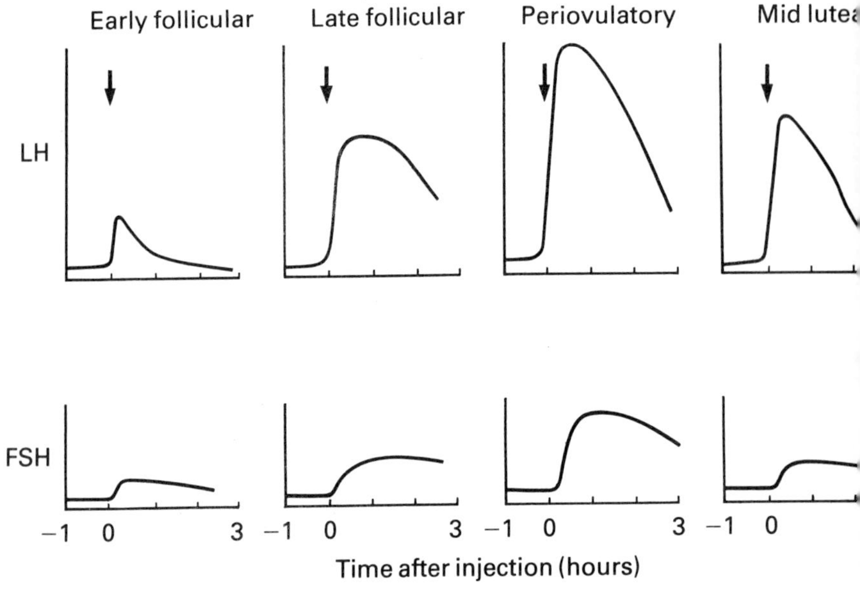

Fig. 1-23. Diagram showing variation in responsiveness to an injection of 100 μg Gn-RH at different stages of the menstrual cycle in women. (Adapted from the data of S. J. Nillius and L. Wide. *J. Obstet. Gynaecol. Brit. Cmwlth.* **79**, 862 (1971) and S. S. C. Yen, *et al. J. Clin. Endocr. Metab.* **35**, 931 (1972).)

and hamsters. This change is largely in response to the increasing levels of oestradiol-17β in the blood, since in rats the effect is abolished if the ovaries are removed or if they are injected with an anti-oestrogen. The effect can be mimicked during the early follicular phase in women by administering oestradiol-17β. Initially, the responsiveness declines because of negative feedback, but after a time responsiveness is considerably enhanced.

These changes in responsiveness are attributed to a direct effect of oestradiol-17β on the sensitivity of the gonadotrophs to Gn-RH but in the normal cycle they should be considered

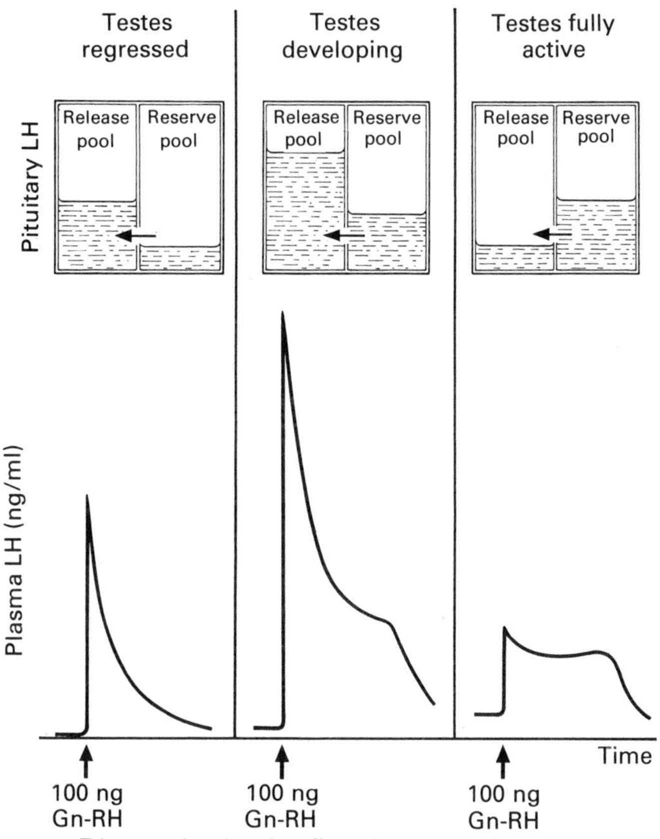

Fig. 1-24. Diagram showing the effect of injection of a constant amount of Gn-RH on secretion of LH, and possible pituitary LH pool sizes at different seasons of the year in the Soay ram. The amplitude of the LH release reflects pituitary responsiveness and the duration of release reflects the size of the reserve pool. (From G. A. Lincoln. *J. Endocrinol.* **69**, 213 (1976).)

together with the changes which oestradiol-17β is having on secretion of Gn-RH. Progesterone, on the other hand, seems to have relatively little effect on pituitary responsiveness in most species. In women, however, when small amounts of progesterone, comparable to the preovulatory increase, are administered during the mid follicular phase of the cycle, there is an enhanced response to Gn-RH.

In the male, changes in pituitary responsiveness to Gn-RH are much less obvious. In the Soay ram, seasonal changes can be demonstrated in pituitary responsiveness to Gn-RH (Fig. 1-24). These may be explained by the changes in testosterone levels (Fig. 1-19) having a direct action on the pituitary, but the frequency of Gn-RH pulses that change during the seasons, as reflected by the number of LH pulses (Fig. 1-19), may be a more important influence.

Changes in pituitary gonadotrophin reserve

An injection of Gn-RH will induce a single release of LH from the pituitary, while infusions of Gn-RH result in a biphasic release in both men and women. The initial discharge can be thought of as coming from a pool of readily releasable LH, and the size of this pool determines the pituitary responsiveness, as described above. The second phase of release evidently represents a reserve pool of recently synthesized hormone. This biphasic pattern can be demonstrated throughout the menstrual cycle in women (Fig. 1-25). Sam Yen and his colleagues in San Diego and David de Kretser and coworkers in Melbourne have found that the size of the second pool increases during the follicular phase. With the approach of ovulation, there is also a substantial increase in the size of the first pool, presumably due to recruitment from the second pool. Because the FSH response to Gn-RH injections is small, and since FSH has a much longer half-life than LH, the precise pattern of FSH release is difficult to analyse. After ovulation, as progesterone begins to be secreted by the developing corpus luteum, the size of the two pools begins to decline again.

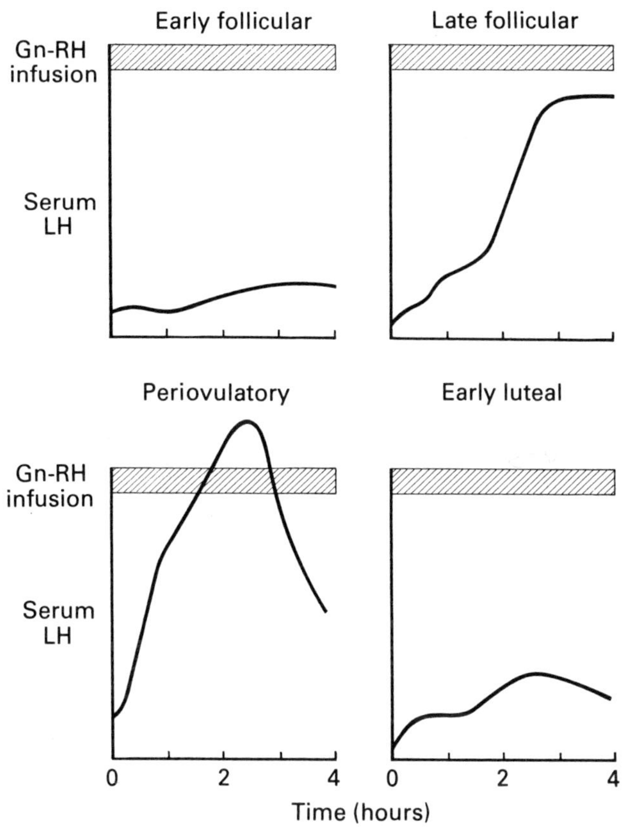

Fig. 1-25. Diagram showing changes in the size of the two pools of LH following infusion of Gn-RH at different stages of the menstrual cycle in women. (Adapted from the data of J. D. Hoff *et al. J. Clin. Endocr. Metab.* **44**, 302 (1977), and D. M. de Kretser *et al. J. Clin. Endocr. Metab.* **46**, 227 (1978).)

In the Soay ram the two-pool pattern of release is apparent after one injection of Gn-RH (Fig. 1-24). The size of the second pool differs during the seasonal cycle, suggesting that synthesis and storage of hormone also varies. Responsiveness is highest in the weeks before maximal testicular activity. This probably involves increased synthesis and storage of LH and acceleration of the mechanism for transfer from the reserve pool to the releasable pool. Ultimately the Gn-RH firing rate may become so high that the releasable pool begins to be depleted.

Self-priming effects of Gn-RH

The existence of a 'self-priming' effect is based on the observation that the response to a second injection of Gn-RH is much greater than the response to the first injection given 1–2 hours previously, indicating a direct action of Gn-RH on the gonadotrophs. The effect is most pronounced just prior to ovulation in rats and women, and may be due to recruitment from the second into the first LH pool in response to the first injection. The effect can be also demonstrated on rat pituitaries *in vitro* but is abolished when inhibitors of RNA and protein synthesis are added to the medium. Whether these inhibitors act directly to inhibit LH synthesis, or whether they act indirectly to prevent synthesis of a protein which is involved in LH release or LH transfer to the releasable pool is not known.

How does hypothalamic activity regulate reproductive function?

We know that several mechanisms are involved in changing the pituitary responsiveness to Gn-RH, but since measuring its secretion from the hypothalamus is extremely difficult we cannot find out directly how hypothalamic activity regulates reproductive function. We have seen that LH is released episodically in a pulsatile manner, so the most obvious assumption is that this is a reflection of a pulsatile release of Gn-RH from the hypothalamus. Thus an increased *frequency* of LH pulses is probably the result of an increased frequency of Gn-RH discharge, whilst the change in *magnitude* of an individual LH pulse is determined by the responsiveness of the anterior pituitary. Although the magnitude of the individual Gn-RH pulses may also change, it is possible that the increased concentration of Gn-RH in the hypophysial portal blood at the time of the LH surge may be due merely to more frequent pulses of Gn-RH occurring during the collection period.

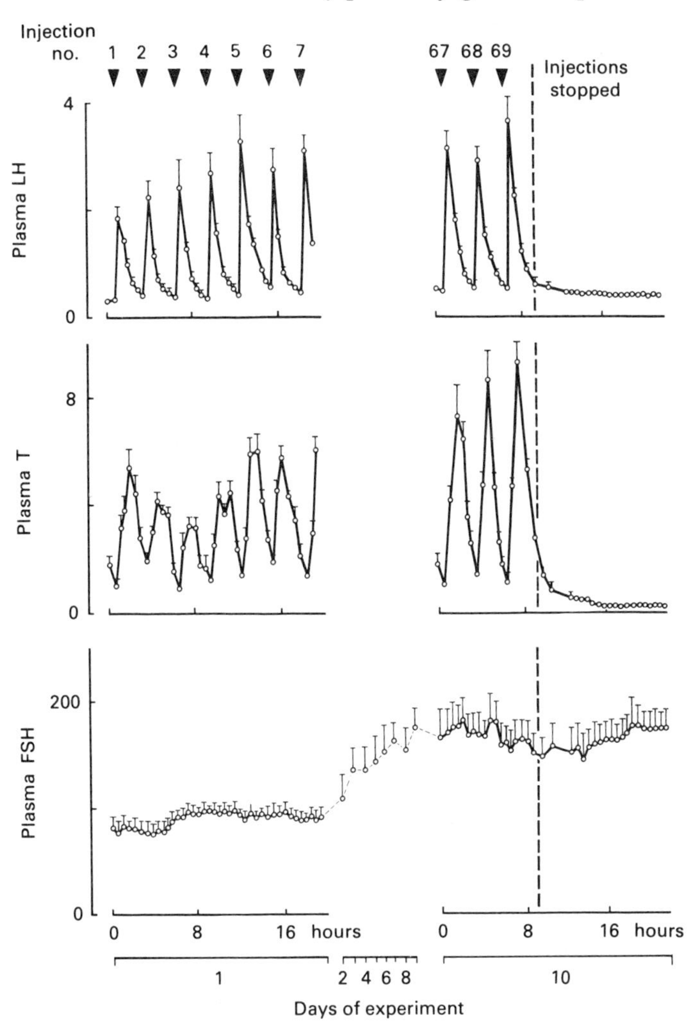

Fig. 1-26. The effect of seven injections of 100 ng Gn-RH each day for 9 days and on the tenth day (indicated by arrows) during the inactive stage of the breeding season in a Soay ram. (G. A. Lincoln. *J. Endocrinol.* **80**, 133 (1979).)

This theory of pulsatile release of Gn-RH can be tested by administering repeated injections of small amounts of Gn-RH at regular intervals, preferably when one would expect the secretion of endogenous Gn-RH to be low. The injection of seventy pulses each of 100 ng Gn-RH over a period of 10 days during the inactive period in the seasonal cycle of the Soay ram produces patterns in secretion of LH, FSH and testosterone remarkably similar to those seen during the resumption of testicular activity in the autumn (Fig. 1–26). A major question in both male and female is how LH and FSH can be stimulated independently by a single releasing hormone which, when injected, releases predominantly LH. The fact that these repeated injections of Gn-RH produce radically different effects on FSH and LH secretion is therefore of particular significance. While each injection of Gn-RH releases the typical pulse of LH, the effect on FSH is minimal at first and it takes many injections before FSH levels rise markedly. Once stimulated, FSH levels remain elevated even 20 hours after the Gn-RH pulses have stopped. Thus, it seems that the changing LH:FSH ratios we see, could result from the different rates at which LH and FSH are being synthesized, stored and released by the gonadotrophs as well as from their different half-lives in the blood.

In the female we have seen how several mechanisms probably act together to cause the preovulatory surge of LH. Thus, the rising levels of oestradiol-17β lead to an increased responsiveness of the pituitary, which is then exposed to an increased amount of Gn-RH. The gonadotrophin available for release increases and the 'self-priming' effect of Gn-RH comes into operation. The mechanisms terminating the LH surge are less clear, but probably result from the decline in the secretion of Gn-RH and a decrease in pituitary responsiveness.

Since the early 1960s, when the first releasing hormone activity was demonstrated in hypothalamic extracts, there has been a tremendous increase in our knowledge about the nature and mode of action of releasing hormones, and already there are synthetic Gn-RH agonists and antagonists available for trial. The 1980s should see the clinical application of this basic research, both in the diagnosis and treatment of infertility, and in the development of new methods of fertility control in man and animals.

Although we know that gonadal steroids are involved in the feedback control of the hypothalamus and anterior pituitary, where there are steroid hormone receptors, we still do not understand precisely how steroids, neutrotransmitters and releasing hormones are related to one another. The releasing hormones are the substances that make the pituitary 'tick', but the central nervous system is unlikely to contain a clock that determines the length of the oestrous or menstrual cycle. In the complete feedback loop of hypothalamus ⟶ pituitary ⟶ gonad ⟶ hypothalamus, the time-lags between stimulus and response in each component part may ultimately decide cycle length in the female. Since the hypothalamus is part of the nervous system, releasing hormone discharge may follow the all-or-none rule for conduction of a nerve impulse, with the system being modulated by changes in frequency of Gn-RH discharges, rather than changes in amplitude. However, a series of complex feedback mechanisms between the pituitary and the gonad ensure that these organs can respond to alterations in the ticking of the hypothalamic metronome by varying both the amplitude, the frequency, and even the ratio of their secretory products. In this way a single signal from the hypothalamus can be translated into a most complex message from the gonads themselves.

SUGGESTED FURTHER READING

FSH-releasing hormone and LH-releasing hormone. A. V. Schally, A. J. Kastin and A. Arimura. *Vitamins and Hormones* **30**, 83 (1972).

Mechanism of action of hypothalamic hormones in the anterior pituitary gland and specific modulation of their activity by sex steroids and thyroid hormones. F. Labrie *et al. Recent Progress in Hormone Research* **34**, 25 (1978).

The Anterior Pituitary. Ed. A. Tixier-Vidal and M. G. Farquhar. London and New York; Academic Press (1975).

Chemistry of hypothalamic hormones. J. Sandow and W. König. In *The Endocrine Hypothalamus*, p. 149. Ed. S. L. Jeffcoate and J. S. M. Hutchinson. London, New York; Academic Press (1978).

Immunofluorescence study of LRF neurones in man. J. Barry. *Cell and Tissue Research* **181**, 1 (1977).

Releasing hormones

Regulation of the hypothalamic-pituitary-ovarian axis in women. S. S. C. Yen. *Journal of Reproduction and Fertility* **51**, 181 (1977).

On the control of gonadotrophin secretion in the rhesus monkey. E. Knobil. *Recent Progress in Hormone Research* **30**, 1 (1974).

Handbook of Physiology Section 7 : Endocrinology, Volume 4, Parts 1 and 2, *The Pituitary Gland and its Neuroendocrine Control*. Ed. E. Knobil and W. H. Sawyer. American Physiological Society (distributed by Williams and Wilkins & Co., Baltimore) (1974).

Neuroendocrine control of gonadotrophin secretion. G. Fink. *British Medical Bulletin*, **35**, 155 (1979).

Photoperiodic control of seasonal breeding in the ram: participation of the cranial sympathetic nervous system. G. A. Lincoln. *Journal of Endocrinology* (1979) in press.

Neuroendocrine control of gonadotrophin secretion in the female rhesus monkey. E. Knobil and T. M. Plant. In *Frontiers in Neuroendocrinology*, vol. 5, p. 249. Ed. W. F. Ganong and L. Martini. New York; Raven Press (1978).

2 Pituitary and placental hormones

Jennifer H. Dorrington

As we have seen in earlier books in this series (notably Book 1, Chapters 2 and 3, and Book 3, Chapter 1) the gonads perform two functions, synthesis of steroid hormones and production of gametes, and both of these processes require the intervention of gonadotrophins. Before discussing the biochemical ways in which the control mechanisms work, we must say something about the biology of these systems.

THE TESTIS

In functional terms, the testis can be thought of as having two regions – the interstitial cell compartment which produces steroids and the seminiferous tubule compartment which manufactures the spermatozoa. The process of spermatogenesis, from spermatogonial stem cell to spermatozoon, takes a long time (74 days in man) and is fascinating in its complexity; the structural changes that take place are known in considerable detail, but our understanding of the biochemical events associated with these changes is still in its infancy. Supporting the germ cells in the seminiferous tubules are the Sertoli cells, which form specialized 'tight junctions' between them so that in essence they make a continuous barrier within every seminiferous tubule (Fig. 2-1). This barrier, referred to as the blood–testis (or lymph–testis) barrier, excludes large molecules such as proteins, and certain other components of the lymph from entering the tubule. Seminiferous tubules transport water from the periphery to the lumen, but this does not occur until tight junctions appear between Sertoli cells. This suggests that Sertoli cells may be involved in creating and maintaining a solute gradient in the adluminal compartment which would cause this movement of water into the tubule.

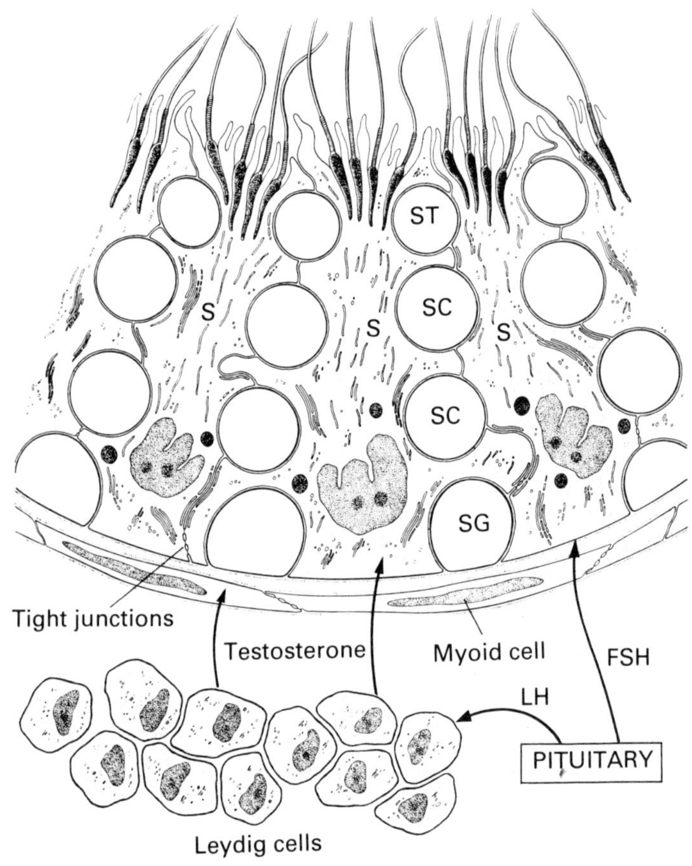

Fig. 2-1. Diagram showing how cells are arranged within the testis. LH acts on Leydig cells, FSH on Sertoli cells and testosterone on myoid cells and Sertoli cells. S, Sertoli cells; SC, spermatocytes; SG, spermatogonia; ST, spermatids.

The blood–testis barrier separates the seminiferous tubule longitudinally into basal and adluminal compartments (Fig. 2-2). In the basal compartment, the spermatogonia undergo mitosis and mature while bathed in a fluid similar to lymph. Primary spermatocytes are formed from the division of type B spermatogonia, and as they proceed through the prophase of meiosis they migrate from the basement membrane and enter the

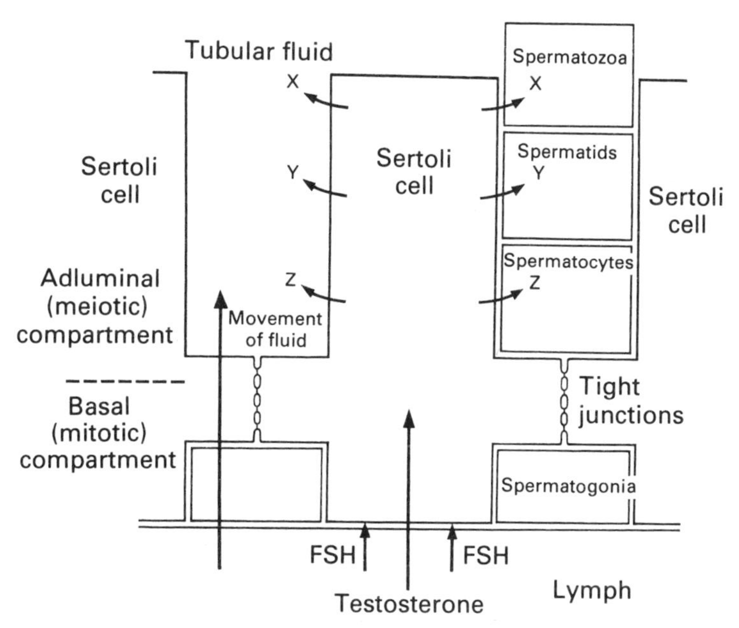

Fig. 2-2. A schematic view of the seminiferous epithelium showing how the tight junctions between Sertoli cells separate the tubule into basal and adluminal compartments, and the transport possibilities for fluid, testosterone and FSH. X, Y and Z represent secretion products of the Sertoli cell.

adluminal compartment. The Sertoli cell secretes proteins, steroids and other components, e.g. inositol, into the adluminal compartment thereby creating a unique environment in which meiosis is completed and haploid spermatids are formed. The subsequent process of spermiogenesis in which spermatids are transformed into spermatozoa involves dramatic structural modifications, and the acquisition of new organelles. This complex metamorphosis can be divided into the Golgi phase, acrosome phase and maturation phase (Book 1, Chapter 3).

Throughout spermiogenesis, the spermatids are in close contact with Sertoli cells and before they break free from their support they discard redundant cytoplasm, packaged in the form of a residual body. The residual body is usually phagocytosed by the Sertoli cell, broken down and presumably re-utilized. The ability of Sertoli cells to ingest not only residual bodies but also degenerating germ cells and other foreign bodies (such as bacteria and particulate matter) has earned them the description of 'macrophage-like'. The spermatozoa are released into the luman and are transported in the stream of tubular fluid to the epididymis.

Removal of the anterior pituitary causes cessation of spermatogenesis and degeneration of germ cells, and we now know that pituitary secretions are required for growth and maturation of the testis, specifically luteinizing hormone (LH) and follicle-stimulating hormone (FSH). Both these gonadotrophins are necessary for the initiation and completion of the first wave of spermatogenesis and also for restoration of sperm production in regressed testes of hypophysectomized rats. Many of the biochemical effects of FSH on the testis decrease or are lost with increasing age. In fact, LH (or testosterone) alone can maintain the whole process of sperm production (albeit at a somewhat reduced yield) in the adult hypophysectomized rat, which certainly suggests that FSH is not needed in the mature animal.

THE OVARY

Like the testis, the ovary can also be divided into different tissue components: the follicles, corpora lutea and interstitial tissue. Shortly after birth, the female gonad contains oocytes that have already proceeded through the early stages of meiosis but have been arrested before its completion. At this stage of development in the female, the ovary contains her life's supply of oocytes (Book 1, Chapter 2). A clear distinction is apparent then between the developmental programme of gametes in the ovary and the testis. In the ovary, oogenesis (the production of oocytes from oogonia) is complete before or shortly after birth, depending upon the species, whereas in the testis there is a continuous proliferation of germ cells and the production of spermatozoa throughout adult life.

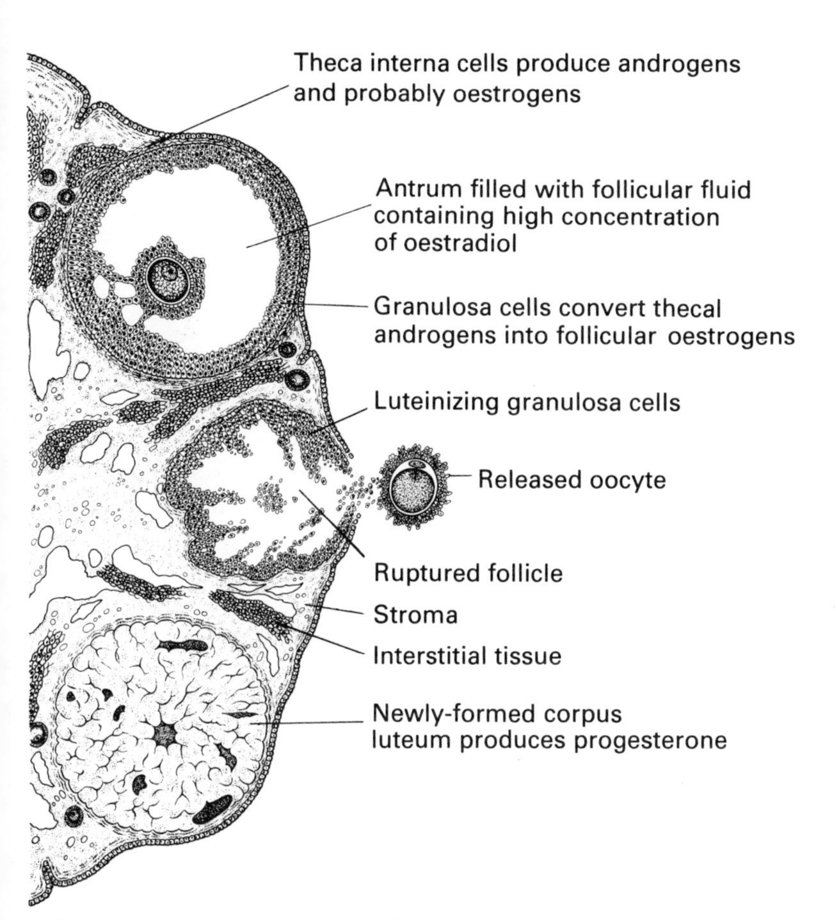

Fig. 2-3. Diagram showing the various cell types in the ovary and their roles in the synthesis of steroids.

While the oocytes are in the resting phase of meiosis (dictyotene) they are surrounded by granulosa cells, forming primordial follicles. The nature of the trigger that initiates growth of a follicle and the reason why a particular follicle is selected are unknown. Granulosa cells in the small growing follicles contain FSH-binding sites, and treatment of neonatal rats with FSH will stimulate follicular growth. At later stages of development, the theca interna and theca externa become organized around the

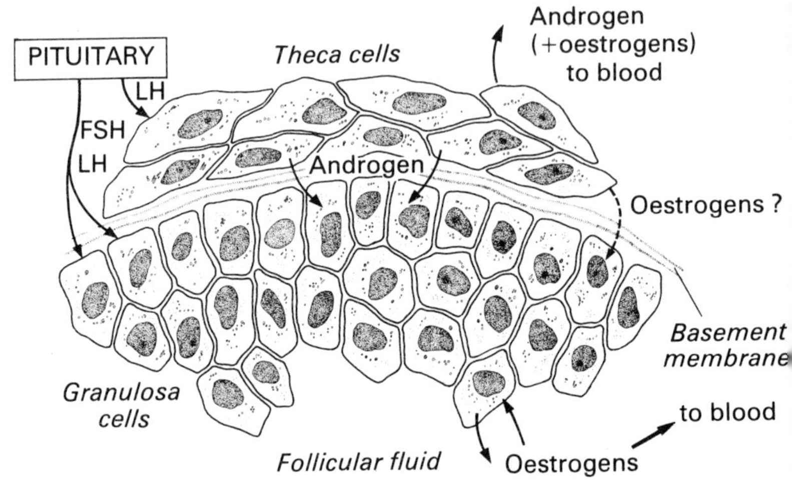

Fig. 2-4. Diagram of part of a follicle showing sites of action of gonadotrophins, and production and action of steroids.

follicles, and follicular fluid accumulates within the antrum (Figs. 2-3 and 2-4). Antrum formation requires the presence of FSH.

Unlike the fluid of the seminiferous tubule, which is rendered virtually free of blood or lymph protein by the restrictions imposed by the blood–testis barrier, follicular fluid resembles serum. Adjacent granulosa cells do not form the 'tight junctions' that are typically seen between Sertoli cells and therefore do not have the capacity to filter out large molecules during the accumulation of follicular fluid. Adjacent granulosa cells interact by forming extensive gap junctions characterized by a network of intercellular pores which facilitate the transfer of small molecules such as 3′,5′-AMP (cyclic AMP) from one cell to another, but do not restrict the movement of proteins in the extracellular space. The thecal and granulosa cells are responsible for the steroid profile and also contribute to the prostaglandin and protein composition of the follicular fluid (Fig. 2-5).

The rapid phase of growth from the antral to the pre-ovulatory stage involves the action of both oestrogen and FSH. During

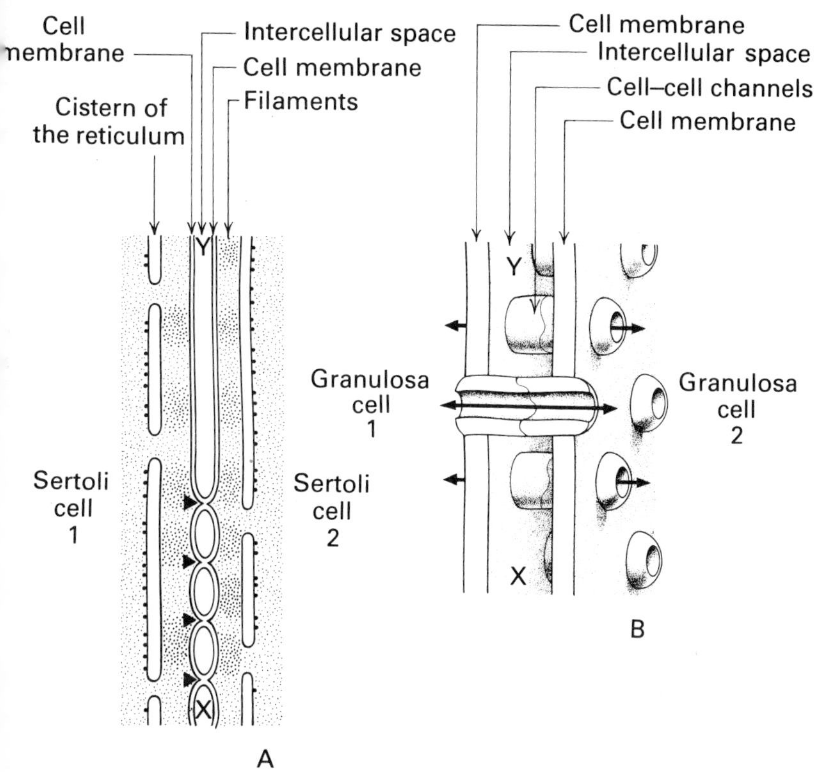

Fig. 2-5. A. Diagram of the 'tight junctions' between two adjacent Sertoli cells. Opposing plasma membranes are fused at multiple sites, as indicated by the arrowheads, and restrict movement of large molecules in the intercellular space from X to Y. (From D. Fawcett. In *Handbook of Physiology*, vol. 5. Ed. R. O. Greep and D. Hamilton. Washington, D.C.; American Physiological Society. (1974).) B. Diagram showing gap junctions which are thought to consist of channels traversing the plasma membranes of two adjacent granulosa cells. The channels would allow passage of ions and small molecules from one cell to another. Large molecules can pass from X to Y in the intercellular space. (From W. R. Loewenstein. In *Cell Membranes : Biochemistry, Cell Biology and Pathology*, Ed. G. Weissman and R. Clayborne. New York; H. P. Publishing (1974).)

this period, FSH induces the appearance of LH-binding sites on granulosa cells in preparation for the dramatic rise in the release of LH from the pituitary. This 'LH-surge' initiates the resumption of meiosis of oocytes in the large pre-ovulatory follicles, the luteinization of granulosa cells to form the corpus luteum, and ovum release. These diverse actions of LH appear to involve different mechanisms. The meiosis-inducing effect of LH on the mammalian oocyte is mediated by cyclic AMP (Fig. 2-9) and requires *de novo* protein synthesis. The effect of LH (or FSH) on progesterone synthesis by granulosa cells also involves cyclic AMP formation, but several lines of evidence indicate that the meiosis-inducing action of LH is not mediated by a steroid.

The LH peak, increased progesterone synthesis and resumption of meiosis take place several hours before follicular rupture and ovum release (for example, the time interval from LH peak to ovulation is 36 hours in women) (Book 1, Chapter 2). Prior to ovum release, part of the follicle wall becomes thinner and weakens to form the stigma, where it ruptures releasing the follicular fluid and the egg. The factors that cause the thinning of the follicle wall have not been clearly defined but may include proteolytic enzymes such as collagenase and plasminogen activator (discussed below). The prostaglandin content of pre-ovulatory follicles increases 4–6 hours after the peak of LH. Dave Armstrong, Hans Lindner and their colleagues have suggested that these biologically active compounds may play an essential role in the events that lead to follicular rupture (Chapter 3). In the presence of indomethacin, which inhibits prostaglandin synthesis, LH induces the resumption of meiosis and stimulates steroid synthesis, but follicular rupture is blocked.

IDENTITY OF PROTEIN HORMONES

The anterior lobe of the pituitary gland secretes a number of polypeptide hormones which influence the metabolism of many different tissues. They include FSH, LH, prolactin, adrenocorticotrophic hormone (ACTH), growth hormone (GH), melanocyte stimulating hormone (MSH) and thyroid stimulating hormone (TSH). Of these, FSH, LH and prolactin interact directly with the ovary and the testis (Book 3, Chapter 1).

The primate placenta produces chorionic gonadotrophin and chorionic somatomammotrophin (Book 3, Chapters 1 and 4). Human chorionic gonadotrophin (hCG) will duplicate the effects of LH on the ovary and the testis, and its mechanism of action appears to be the same as that of LH. Both hCG and human chorionic somatomammotrophin (hCS) are probably involved in the maintenance of the corpus luteum of pregnancy. hCG may be important in regulating the endocrine activity of Leydig cells in the fetal testis. Leydig cells undergo two phases of development, the first taking place in the fetus and the second at puberty. In the fetal wave, the Leydig cells secrete testosterone which is essential for the differentiation of the male reproductive tract (Wolffian ducts), whilst the Sertoli cells produce 'Mullerian duct inhibiting factor' which causes the regression of the Mullerian ducts (the embryonic female reproductive tract).

STRUCTURE OF GONADOTROPHINS

LH, hCG and FSH are glycoproteins, the protein core having branched carbohydrate side-chains. Their molecular weights are in the 30000–35000 range. These hormones consist of two dissimilar subunits designated α and β, which are non-covalently linked (Fig. 2-6). The amino acid sequences of the α-subunits of FSH, LH, CG and TSH are essentially similar, whereas β-subunits are unique and undoubtedly account for hormonal specificity. Prolactin on the other hand is a single polypeptide chain with a molecular weight of 24000. These biologically active molecules apparently do not enter the cell to exert their effects but interact with specific recognition sites or receptors on the surface of the target cell (Book 3, Chapter 1).

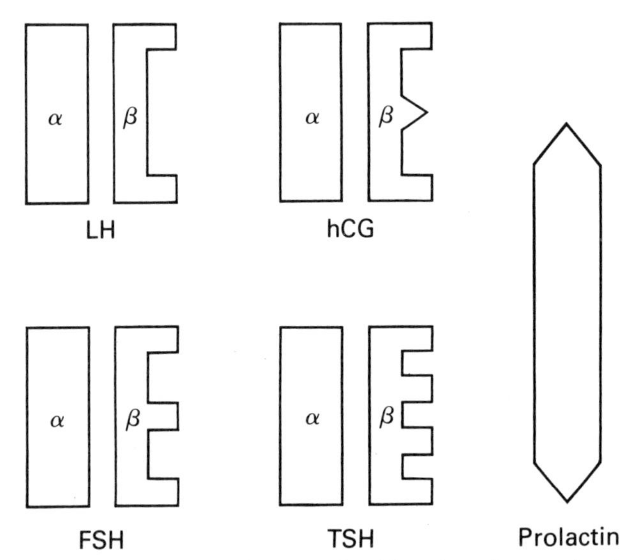

Fig. 2-6. A diagram to illustrate the idea that the α-subunits of LH, hCG, FSH and TSH are structurally homologous whereas the β-subunits are specific for each hormone. In contrast, prolactin is thought of as a single polypeptide chain.

TARGET CELLS FOR GONADOTROPHINS

Target cells for FSH in the testis

In the rat testis, only the seminiferous tubules bind FSH. Of the cell types present in the tubule, i.e. Sertoli, myoid and germ cells, only Sertoli cells bind radioactively-labelled FSH specifically, and the bulk of the radioactivity is associated with the plasma membrane fraction. Treatment of cells with proteolytic enzymes destroys the FSH-binding activity, suggesting that the hormone receptor may be protein.

Binding of a hormone can be considered to be a primary event in the action of that hormone if the interaction causes a measurable response. Clearly, it was important to see if FSH could modulate Sertoli cell function in any way. This has proved

TABLE 2-1. Summary of the responses of gonadal support cells to the action of FSH

Stimulated response	Testicular Sertoli cells	Ovarian granulosa cells
Cyclic AMP accumulation	Yes	Yes
Protein kinase activation	Yes	Yes
DNA synthesis	Yes	Yes
Protein synthesis	Yes	Yes
Plasminogen activator activation	Yes	Yes
ABP synthesis	Yes	Not known
LH binding	No	Yes
Progesterone synthesis	No	Yes
Oestradiol synthesis	Yes	Yes

to be the case, and a number of end-responses of Sertoli cells to the action of FSH have now been defined (Table 2-1).

FSH and testosterone stimulate Sertoli cells to synthesize and secrete androgen-binding protein (ABP). FSH also stimulates Sertoli cells to secrete plasminogen activator. ABP is a unique product of the Sertoli cell and binds the steroids testosterone and dihydrotestosterone with a high affinity. The ABP is transported from the Sertoli cell through the seminiferous tubules and rete testis to the epididymis, where it is absorbed. The physiological role of this binding protein in the testis and epididymis is not clear, but whatever its function may be we have to acknowledge the sobering fact that the boar is quite fertile without it. Plasminogen activator has the function of converting plasminogen (present in serum and lymph) to plasmin, which is a proteolytic enzyme. We can speculate that the release of plasminogen activator and the subsequent production of plasmin may be involved in the opening of the tight junctions between Sertoli cells to allow the movement of germ cells from the basal to the adluminal compartment. Another possibility is that plasminogen activator is required for the release of spermatozoa

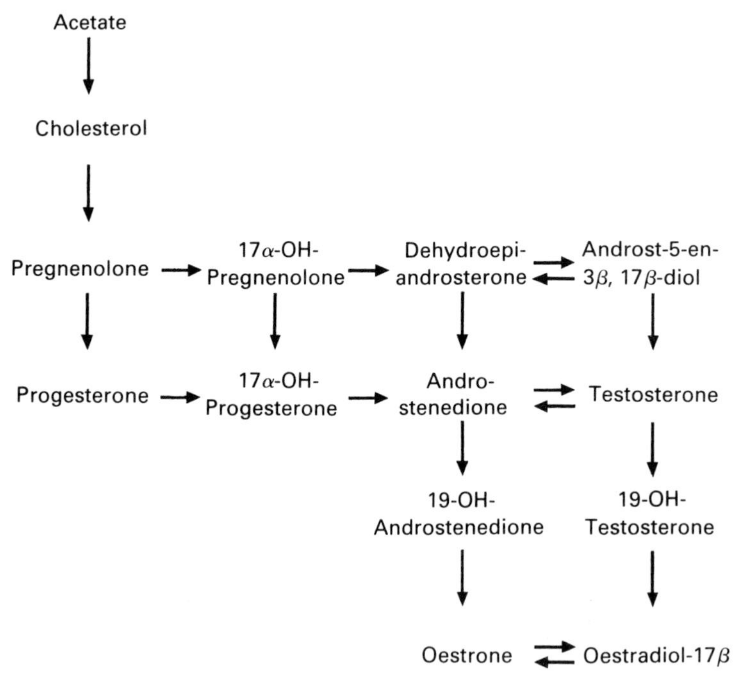

Fig. 2-7. Pathways of biosynthesis of progesterone, testosterone and oestrogens from acetate and cholesterol in the testis.

into the lumen of the seminiferous tubule. Other cells that secrete plasminogen activator include activated macrophages, ovarian granulosa cells (see below) and many tumour cells, and in general its production is associated with cells that are restructuring on a basement membrane.

The Sertoli cell cannot synthesize androgens from cholesterol, and relies upon the Leydig cell for its source of C-19 steroids (Fig. 2-7). However, immature Sertoli cells can convert testosterone to oestradiol-17β and oestrone, when stimulated by FSH (Fig. 2-8). The Sertoli cell and Leydig cell are the only cell types in the testis that have the capacity to synthesize oestrogens.

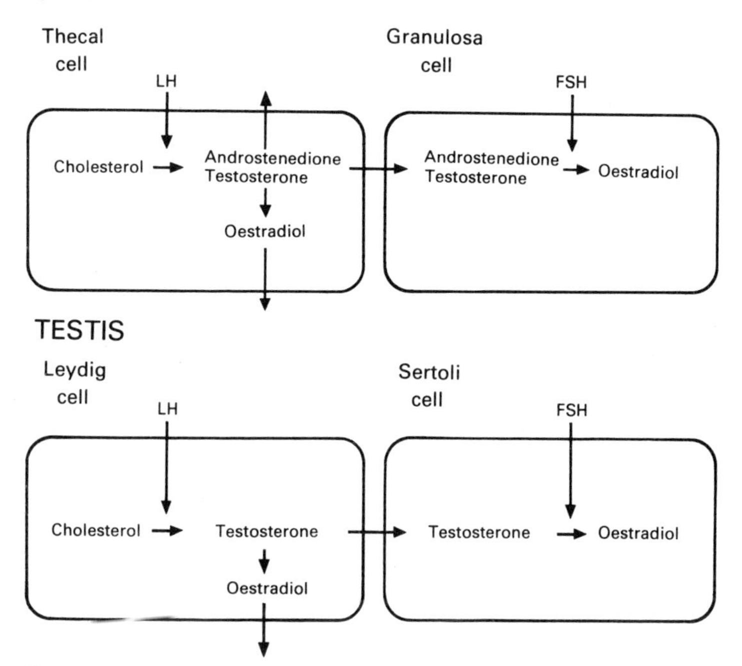

Fig. 2-8. Two-gonadotrophin–two-cell-type model for the synthesis of oestradiol in the gonads.

Target cells for LH in the testis

The primary target for LH in the testis is the Leydig cell (Table 2-2). Leydig cells possess cell surface receptors that bind labelled LH specifically, whereas the cell types present in the seminiferous tubule will not bind LH. This may seem incongruous at first since LH is required for spermatogenesis to proceed but it can be explained by the finding that testosterone can completely replace LH in supporting spermatogenesis. Thus it appears that the primary function of LH is to regulate the production of testosterone which is either exported via the blood stream to other tissues or passes into the lymph that bathes the seminiferous tubules (Fig. 2-1). Because of the slow flow of lymph, a high concentration of testosterone is maintained in the vicinity of the tubule. Testosterone binds specifically to peritubular myoid cells and Sertoli cells, but not to germ cells. Since testosterone does not interact directly with germ cells, there is still a missing link, because androgen action is required for the completion of meiosis in these cells. The identification of the trigger for the completion of meiosis, presumably emanating from the Sertoli cells under the control of testosterone, will constitute a landmark in future research.

Target cells for FSH in the ovary and testis

FSH binds only to the granulosa cells in the mammalian ovary. Granulosa cells in some large follicles bind labelled FSH and labelled hCG whereas those in small and medium sized follicles bind only FSH. In the ovary and the testis, therefore, FSH influences only one cell type directly, the granulosa cell or the Sertoli cell. FSH appears to be involved in the initiation of fluid movement into the follicles and seminiferous tubules; granulosa cells, provide them with nutrients and establish a microenvironment that is conducive to germ cell development (Table 2-1). products to a plasma ultrafiltrate before delivering it to the lumen of the seminiferous tubules. These fluids bathe the germ cells, provide them with nutrients, and establish a microenvironment that is conducive to germ cell development (Table 2-1).

In addition to stimulating the passage of follicular fluid into the antrum, FSH also stimulates protein synthesis and increases the mitotic activity of granulosa cells in growing follicles. Oestrogen is a more effective mitogen for granulosa cells than FSH, and causes increases in ovarian weight when injected into immature rats or when applied topically. Treatment with oestrogen in conjunction with FSH produces larger follicles than does FSH alone, since both the mitotic activity of granulosa cells and the accumulation of fluid is enhanced. FSH acts on the follicle to promote growth and maturation. One of the most important 'maturational' functions of FSH in the large follicle is to increase the number of LH binding sites on granulosa cells. The mechanism by which FSH effects this change is not known, but may involve either activation of pre-existing proteins or induction of protein synthesis. Granulosa cells from pre-ovulatory follicles therefore contain receptors for both FSH and LH on their cell surfaces. As shown by Cornelia Channing, both FSH and LH increase the level of cyclic AMP within granulosa cells, and this nucleotide is believed to be the intracellular mediator for both gonadotrophins. This common mechanism of action may explain the intrinsic ability of FSH to mimic the effects of LH on follicles, namely progesterone accumulation, the resumption of meiosis and induction of ovulation. Bill Beers has found that FSH is more effective than LH in stimulating granulosa cells to synthesize plasminogen activator. Granulosa cells *in vivo* produce increasing amounts of plasminogen activator as the time of ovulation approaches, and we can suppose that plasminogen activator is involved in the breakdown of the follicle wall, which precedes ovum release. Plasminogen, the substrate for plasminogen activator, is present in follicular fluid, and the product of the reaction, plasmin, has been shown to weaken strips of follicle wall *in vitro*.

Granulosa cells isolated from a number of mammalian species (e.g. rat, man, pig and horse) have the capacity to convert cholesterol to progesterone but are deficient in the enzymes that convert progesterone to androgens (Fig. 2-7). Consequently, granulosa cells rely upon the thecal and interstitial cells to provide them with adequate supplies of androstenedione or testosterone. Granulosa cells contain an aromatization enzyme system which converts androgens to oestrogens, and this process is stimulated by FSH. The co-operation of two cell types and two gonadotrophins is apparently required for oestrogen biosynthesis to take place within the follicle or seminiferous tubule. The thecal cells and Leydig cells, when stimulated by LH, synthesize androstenedione and testosterone which are transported to the granulosa cells and Sertoli cells, for aromatization to oestrogens under the influence of FSH (Fig. 2-8).

Target cells for LH in ovary and testis

The tissue components of the ovary – the interstitial tissue, follicles, and corpora lutea – synthesize characteristic profiles of steroids, and in each case this is influenced by LH (Table 2-2).

TABLE 2-2. Direct effects of LH on cells in gonads

Gonad	End-response
Testis	
Leydig cells	Increased steroidogenesis (androgens +oestrogens)
Sertoli cells	None
Myoid cells	None
Germ cells	None
Ovary	
Interstitial cells	Increased steroidogenesis (androgens, gestagens)
Theca interna cells	Increased steroidogenesis (androgens and oestrogens)
Granulosa cells	Increased steroidogenesis (progesterone and oestrogen)
Corpus luteum cells	Increased steroidogenesis (progesterone)
Oocyte	None

The follicle synthesizes three classes of steroids: namely, the progestational hormones, androgens and oestrogens. The relative proportions of these depend upon the stage of development of the follicle. Even though the synthesis of steroids is a well-recognized function of the follicle, there is still some controversy over the roles played by the theca interna and granulosa cells. Thecal cells have been isolated by microdissection from the follicles of several species (e.g. man, rabbit, rat and hamster) and have the capacity to synthesize androgens, and probably also oestrogens in some species. LH binds to thecal cells and is responsible for the control of steroidogenesis. In man, monkey, horse and sheep, androstenedione produced by the thecal cells is quantitatively the major steroid secreted by the ovary just before ovulation, and testosterone is produced only in small amounts. The androstenedione or testosterone produced by the thecal tissue can pass into the circulation or it can diffuse into the granulosa cell layer or other parts of the ovary. That granulosa cells are target cells for androgen action is indicated by Griff Ross's discovery that they contain specific androgen receptors, and this can be inferred also from the ability of testosterone to induce a measurable response, namely, elevated progesterone synthesis. Progesterone synthesis is greatly enhanced when granulosa cells are exposed simultaneously to FSH (or LH in the case of cells from large follicles) and testosterone. How the steroid hormone and the polypeptide hormone interact to bring about this effect is intriguing, but little is known of the mechanisms involved. In addition to its hormonal influences on granulosa cells, testosterone can serve as a substrate for the synthesis of 5α-reduced androgens (dihydrotestosterone and androstanediols) and oestrogens.

Following the LH-induced resumption of meiosis of the oocyte and ovulation, the follicle, which hitherto secreted mainly oestrogen and androgen, is transformed into a corpus luteum which synthesizes large amounts of progesterone. This process is called luteinization and is also initiated by LH.

Once ovulation and luteinization have been induced by LH, striking species differences are evident in the hormonal requirements for the maintenance of a functional corpus luteum. LH is the major luteotrophin in some animals (e.g. the cow), while in others LH constitutes a component of the 'luteotrophic complex' which also includes prolactin and oestrogen.

'Down' regulation of receptors

Certain drugs and hormones become less effective after repeated administration. Recently, high concentrations of hormones have been found to cause negative or 'down' regulation of the production of their own receptors. For example, elevated levels of insulin cause a decrease in insulin receptors in target tissues. Similarly in the gonads, Kevin Catt and colleagues in the United States and Richard Sharpe in Scotland have shown that in addition to the effects of LH on steroid production this hormone can also regulate the concentration of its own receptor on the surface of target cells. Exposure of animals to high concentrations of LH *in vivo* causes within hours a parallel loss of LH-binding capacity and stimulation of adenylate cyclase activity in both ovarian cells and Leydig cells. The normal complement of receptors is not recovered for several days. The physiological significance of this phenomenon in the male is not clear. In the female, the loss of LH receptors and adenylate cyclase activity may be important in LH-induced luteolysis and also may help to explain the decline in oestrogen secretion by the Graafian follicles following the pre-ovulatory LH surge.

PROLACTIN

Prolactin is best known for its effects on the mammary gland and on lactation, but must also be included as a gonadotrophin because it has been shown to play a definite role in the regulation of luteal function in the sheep, rat, mouse and probably man. Prolactin levels reach a peak during the oestrous cycle at about the same time as the LH surge that induces ovulation. These elevated amounts of prolactin may assist in the luteolysis of the previous crop of corpora lutea. As suggested by the work of Ken McNatty and his colleagues, prolactin may have a double threshold action on the human corpus luteum, since low levels are required for normal progesterone production, whereas high concentrations are inhibitory. After ovulation, the newly-formed corpora lutea acquire increased numbers of LH receptors and secrete progesterone in response to prolactin. Prolactin is essential during early pregnancy and plays a role together with LH in maintaining progesterone secretion.

MECHANISM OF ACTION OF GONADOTROPHINS

Role of cyclic AMP

Increased intracellular formation of 3′,5′-AMP (cyclic AMP) is thought to be a general mechanism by which a variety of protein hormones produce their biochemical effects. These hormones include FSH, LH, hCG, TSH, ACTH, insulin, vasopressin and

Adenosine 3′, 5′-monophosphate

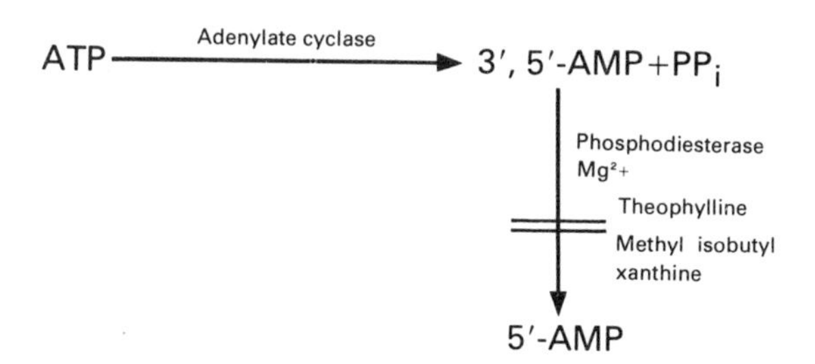

Fig. 2-9. The structural formula of cyclic AMP and the reactions catalysed by adenylate cyclase and phosphodiesterase.

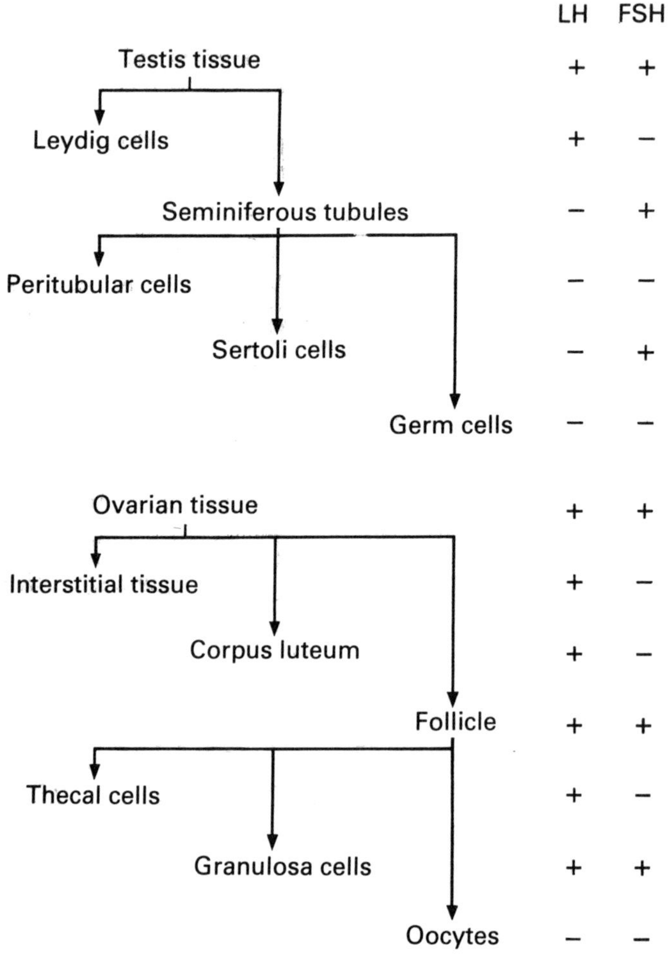

Fig. 2-10. Ability of LH and FSH to stimulate cyclic AMP production by cell preparations from the gonads.

glucagon. Binding of these hormones to the specific receptors on the plasma membranes of target cells causes the activation of adenylate cyclase, an enzyme located on the inner surface of the plasma membrane (Fig. 2-9).

A great deal of work has implicated cyclic AMP as the

intracellular mediator (or 'second messenger') of many actions of LH and FSH in the ovary and the testis (Fig. 2-10). Evidence that is consistent with this role in gonadal preparations falls into three categories: (1) the addition of hormones to the target tissue (or cells) results in a rapid stimulation of cyclic AMP production, which precedes increases in the end-responses; (2) the addition of cyclic AMP (or a derivative such as dibutyryl 3′,5′-AMP) to the tissue *in vitro* will mimic the effect of the hormone on the end-response; (3) agents like theophylline and methyl isobutyl xanthine, which inhibit phosphodiesterase, will also enhance the end-response (Fig. 2-9).

Protein kinase activation

The many diverse responses produced by cyclic AMP in different tissues may involve the activation of protein kinases. Cyclic-AMP-dependent protein kinases are composed of regulatory subunits (R) and catalytic subunits (C). The binding of the regulatory subunits to the catalytic component gives an essentially inactive enzyme. Cyclic AMP promotes the dissociation to yield a regulatory [subunit + 3′,5′-AMP] complex and a free enzymatically active, catalytic subunit.

$$\text{R–C (inactive)} + 3',5'\text{-AMP} \rightarrow \text{R–}3',5'\text{-AMP} + \text{C (active)}$$

There is a widespread distribution of cyclic-AMP-dependent protein kinases in mammalian tissues. Protein kinases transfer the phosphate from ATP to a substrate and consequently by this process of phosphorylation, the enzymes could activate proteins important in the control of cell function.

Mechanism of action of FSH

Compared to the other glycoprotein hormones little is known of the mechanism of action of FSH. From the information available, one can propose the following sequence of events: FSH binds to its target cells leading to the activation of adenylate cyclase,

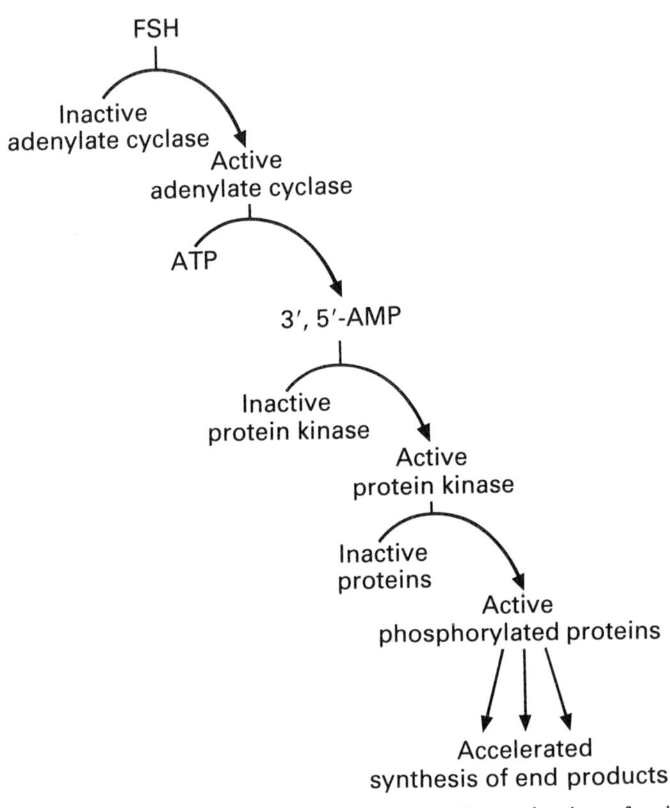

Fig. 2-11. Cascade series of events involved in the mechanism of action of FSH.

which results in increases in the intracellular concentration of cyclic AMP. Cyclic AMP then activates protein kinases resulting in the phosphorylation of proteins, which are essential links in the chain of events. The phosphorylated proteins may bring about their effects in three main ways: (i) by influencing the transcription of DNA, and thus the synthesis of specific messenger RNAs which code for proteins required for the expression of the FSH effects; (ii) by influencing cell functions at the translational level, and thus protein synthesis; or (iii) by acting directly on the rate-limiting steps in the reactions involved (Fig. 2-11).

Of the known effects of FSH on cellular functions (Table 2-1), the stimulatory action on oestradiol synthesis has been studied in the greatest detail. Cyclic AMP derivatives could produce the same effects as FSH, both qualitatively and quantitatively. There was a time lag before oestradiol synthesis was stimulated by either FSH or cyclic AMP derivatives. Studies with inhibitors of RNA synthesis (actinomycin D) and protein synthesis (puromycin) suggested that new messenger RNA synthesis and subsequent translation to provide a unique protein product was taking place during this time lag and was essential for the manifestation of the FSH effect.

Mechanism of action of LH (and hCG) on steroidogenesis

Of the effects of gonadotrophins on the ovary and the testis the stimulation of steroidogenesis by LH has been studied most thoroughly, but even so its mechanism is far from being well understood.

The synthesis of progesterone from cholesterol proceeds by the same pathway in the adrenal, the ovary and the testis. LH and ACTH accelerate steroidogenesis in their respective target tissues by influencing the conversion of cholesterol to pregnenolone, which is the rate-limiting step in the biosynthetic pathway (Fig. 2-12).

The possible involvement of cyclic AMP in the regulation of steroidogenesis was first indicated in 1958 when the steroidogenic response to ACTH was found to be associated with elevated levels of cyclic AMP. Exogenous cyclic AMP was shown to stimulate steroid synthesis in isolated rat adrenals and it was proposed that cyclic AMP mediated the steroidogenic effect of ACTH. Guided by this work on the adrenal, other investigators carried out similar studies on the ovary and the testis and showed that cyclic AMP could also stimulate steroido-

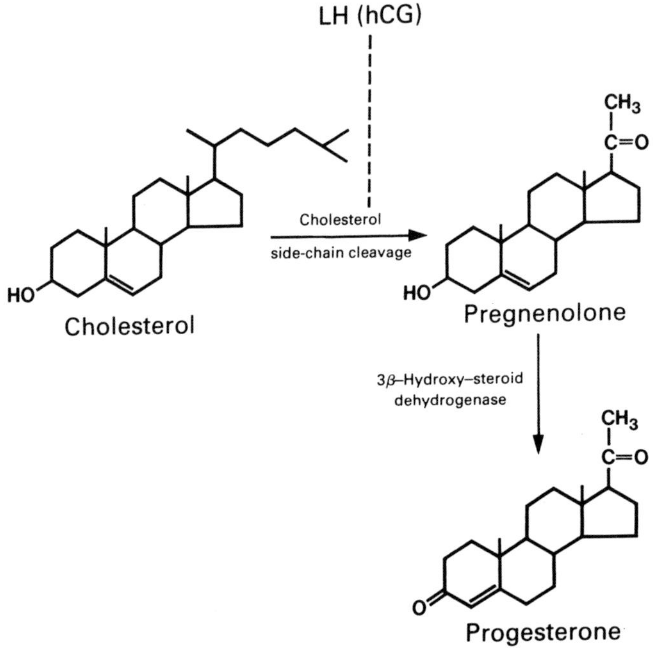

Fig. 2-12. Reactions involved in the conversion of cholesterol to progesterone.

genesis in those tissues. LH increased the level of cyclic AMP in gonadal tissues rapidly and this effect preceded the increase of steroidogenesis.

If we are convinced by the considerable body of evidence that cyclic AMP is a mediator of LH action on steroidogenesis, we are still confronted with the problem of how this nucleotide can accelerate the conversion of cholesterol to pregnenolone. There are several possible explanations: cyclic AMP may accelerate this reaction by increasing (i) the level of the cofactor NADPH, (ii) the total amount of intracellular free cholesterol, (iii) the transport of cholesterol to the enzyme system in the mitochondrion responsible for cleavage of the cholesterol side-chain, or (iv) the activity of the side-chain cleavage enzyme system. Evidence has been presented for and against each of the pos-

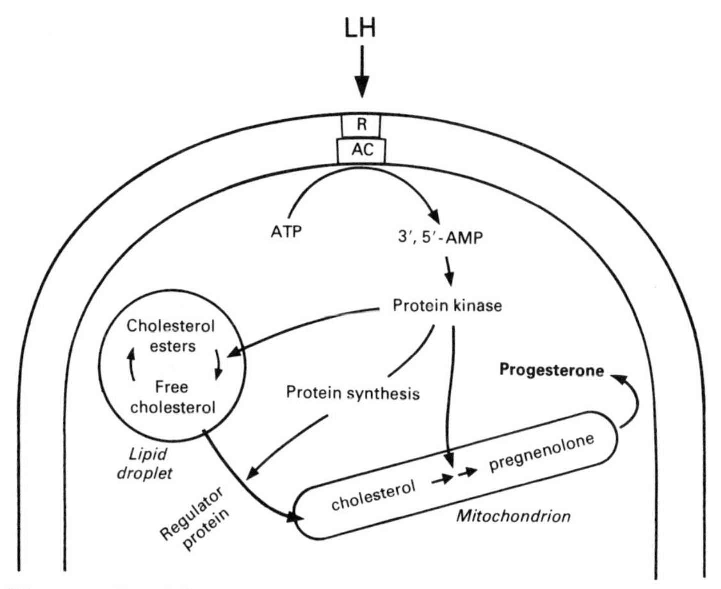

Fig. 2-13. Possible sites of action of LH and cyclic AMP on steroidogenesis. R, the hormone receptor; AC, adenylate cyclase.

sibilities, but the concensus of opinion at this time is that the primary action of LH (and cyclic AMP) is on the transport of cholesterol into the mitochondrion.

Puromycin and cycloheximide inhibit the effects of both LH and ACTH on steroidogenesis in their target tissues. The existence of a regulator protein has been proposed to explain the effects of these inhibitors of protein synthesis. In the ovary, testis, and adrenal, this hypothetical protein is thought to exert its effects before cholesterol side-chain cleavage, possibly at the level of cholesterol transport to the catalytic site (Fig. 2-13). Further studies on this putative protein, and the mechanism by which cholesterol is translocated, could provide new insight into the control of steroidogenesis.

One of the early landmarks in reproductive biology was the demonstration that gonadal functions require the presence of the same hormones, FSH and LH, in the male and the female. We know now that these gonadotrophins act on the somatic cells of the gonads and that there are many similarities in the mechanisms by which they exert their effects.

After many years of controversy, FSH has now been clearly shown to bind specifically to Sertoli cells and granulosa cells, where it stimulates cyclic AMP production, protein kinase activation and protein synthesis. There is little doubt that FSH is required for testicular maturation and for the initiation of spermatogenesis. Many of the effects of FSH in the testis are lost with increasing age of the animal, and even though FSH continues to be synthesized and released by the pituitary throughout adult life, and the Sertoli cells retain FSH receptors, the role of this gonadotrophin in the mature animal is obscure.

Testosterone is essential for fertility in the adult male. We now know the biosynthetic pathway by which testosterone is synthesized in the Leydig cells and the rate-limiting step in this pathway that is influenced by LH. LH accelerates testosterone synthesis by triggering a number of chemical reactions including cyclic AMP production, protein kinase activation and protein synthesis. Not only is the testis the site of production of testosterone but is also its principal target organ. Testosterone influences the differentiation of myoid cells and modulates Sertoli cell function but does not act directly upon the germ cells. These represent significant advances in the understanding of the control of spermatogenesis, but many questions remain unanswered, not the least of which is the nature of the products secreted by the Sertoli cell under the influence of FSH and testosterone, which are crucial for germ cell differentiation.

Over 30 years ago, Roy Greep showed that the transformation of a growing follicle into an antral follicle which secretes oestrogen required both FSH and LH. LH acts on the thecal cells to stimulate the production of androgens which in turn influence granulosa cell metabolism, and also serve as substrates (androstenedione and testosterone) for oestrogen biosynthesis. Oestrogens stimulate the mitotic activity of granulosa cells during the rapid phase of growth of antral follicles to the pre-ovulatory stage. FSH is also required for the growth of ovarian follicles to the point at which they are full of fluid and bulging from the surface of the ovary. During this period of preparation for ovulation, FSH 'turns on' LH receptors in readiness for the LH surge.

We have begun to understand the hormonal control of the somatic elements of the gonads and we know something of the roles of the gonadotrophins and sex steroids in the production of the gametes. But mystery still surrounds the mechanisms permitting oogenesis to begin before or shortly after birth while restraining spermatogenesis from starting until puberty.

Anne Grete Byskov in Copenhagen was the first to show that some secretion of the rete ovarii initiated meiosis in the fetal ovary, while Alex Tsafriri and Cornelia Channing demonstrated that components of follicular fluid – as yet uncharacterized – inhibit the resumption of meiosis.

The gonadotrophins, LH, FSH and prolactin, and steroids testosterone and oestrogens, all interact with the somatic cells of the gonads. Upon receiving the appropriate stimuli the somatic cells produce factors that allow the formation of the male and female gametes. Having begun to understand the hormonal control of the somatic elements of the gonad, the challenge for biochemists in the future is to discover how the gonadal and gonadotrophic hormones regulate meiosis.

SUGGESTED FURTHER READING

Role of follicle-stimulating hormone and luteinizing hormone in follicular development and ovulation. G. S. Greenwald. In *Handbook of Physiology, Endocrinology*, vol. 4, part 2, chapter 33, p. 293. Ed. R. O. Greep and D. Hamilton. Washington, D.C.; American Physiological Society (1975).

Ultrastructure and function of the Sertoli cell. D. W. Fawcett. In *Handbook of Physiology, Endocrinology*, vol. 5, part 2, chapter 2, p. 21. R. O. Greep and D. Hamilton. Washington, D.C.; American Physiological Society (1975).

Pituitary and placental hormones

Gonadotrophin action on cultured Graafian follicles: induction of maturation division of the mammalian oocyte and differentiation of the luteal cell. H. R. Lindner, A. Tsafriri, M. E. Lieberman, U. Zor, Y. Koch, S. Bauminger and A. Barnea. *Recent Progress in Hormone Research* **30**, 709 (1974).

Protein hormone action: a key to understanding ovarian follicular and luteal cell development. J. S. Richards and A. R. Midgley. *Biology of Reproduction* **14**, 82 (1976).

Control of testicular estrogen synthesis. J. H. Dorrington, I. B. Fritz and D. T. Armstrong, *Biology of Reproduction* **18**, 55 (1978).

The role of cyclic AMP in gonadal function. J. M. Marsh. *Advances in Cyclic Nucleotide Research* **6**, 137 (1975).

3 Prostaglandins

John R. G. Challis

The story of the prostaglandins began very simply. In 1930, Kurzrok and Lieb of New York City showed that fresh human semen could make strips of human uterus relax or contract, depending upon whether the patient from whom the organ had been removed had previously borne children or was sterile. Three years later, other workers found that if extracts from human seminal fluid were injected into animals, the blood pressure was lowered, and at the same time, smooth muscle was stimulated to contract. Ulf von Euler at the Karolinska Institute in Stockholm noticed that the same effects could be produced by extracts of seminal fluid from other animals, and he called the active agent 'prostaglandin' because he thought it came from the prostate gland. The prostate is not in fact a major source; prostaglandins – there are actually a number of related substances – come from the seminal vesicles and are also formed in many other tissues.

These early observations evoked very little enthusiasm from the scientific community, and most people seemed prepared to think of the prostaglandins as mere chemical oddities. Quite suddenly, in the early sixties, the scene changed dramatically and in the last 10–15 years we have seen an information explosion on the prostaglandins. This was led by the pioneering work on structure by Sune Bergstrom and Bengt Samuelsson in Sweden, followed by the detailed description of the large and potent prostaglandin family, members of which can have varied and often opposing (balancing) influences. The family is still growing rapidly. In 1973, Samuelsson's group described the addition of a new branch, the thromboxanes, and in 1976 Vane's group in England introduced a further unstable but highly potent relative, prostacyclin.

PROSTAGLANDIN BIOSYNTHESIS

Prostaglandins differ from the classical hormones. Their concentrations in the circulating blood are very low, particularly in comparison to steroid or peptide hormones. In most tissues, they function as highly potent local agents, effective near to or at their sites of production. The rest of the body is 'protected' by the rapid rate of prostaglandin breakdown in the lung and the liver.

If prostaglandins are to function as local hormones, factors

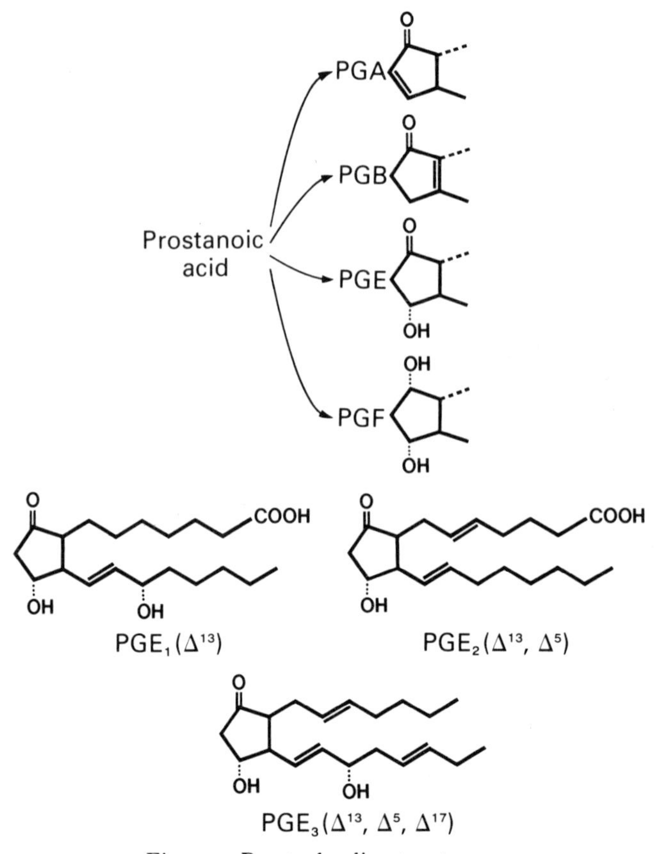

Fig. 3-1. Prostaglandin structures.

that influence their local concentrations are of considerable importance in regulating their activity. In general, these compounds are not stored at the site of production but are synthesized and released rapidly in response to an appropriate stimulus. And so we must look closely at the key aspects of their biosynthesis before trying to evaluate their many roles in reproductive processes.

Prostaglandins are 20-carbon carboxylic fatty acids (see Book 3, Chapter 1) consisting of two side-chains joined by a cyclopentane ring structure. They are synthesized in the cell microsomes from essential fatty acids. Prostaglandins are structural derivatives of prostanoic acid and are classified into four main groups, A, B, E and F, which differ in the ring substituents and the number of double bonds in the molecule (Fig. 3-1). For example, prostaglandin E_1 (PGE_1) has a double bond between carbon atoms 13 and 14, referred to as Δ^{13}; PGE_2 has Δ^5 and Δ^{13} and PGE_3 has Δ^5, Δ^{13} and Δ^{17}.

A major precursor for prostaglandin biosynthesis is cell membrane phospholipid. PGE_2 and $PGF_{2\alpha}$ production depends on the hydrolysis of esterified arachidonic acid (from membrane phospholipid) which yields unesterified arachidonic acid. Arachidonic acid is present largely in the 2-position of the phospholipid molecule and the activity of the enzyme responsible for its cleavage, phospholipase A_2, may control a critical regulatory step in many prostaglandin-generating systems. Whilst phospholipids are a major source of the bonded form of the compound (arachidonate), free arachidonic acid is also present in blood plasma though it comprises only 1–2 per cent of the total plasma free fatty acid.

Free arachidonic acid is metabolized through a fatty acid cyclo-oxygenase enzyme system to form the unstable endoperoxides, PGG_2 and PGH_2 (with a half-life in saline of approximately 5 minutes). The endoperoxides may be converted by an isomerase and peroxidase to PGE_2, or reduced non-enzymatically to $PGF_{2\alpha}$ (Figs 3-2 and 3-3). Arachidonic acid may also be metabolized to other highly unstable compounds, thromboxane

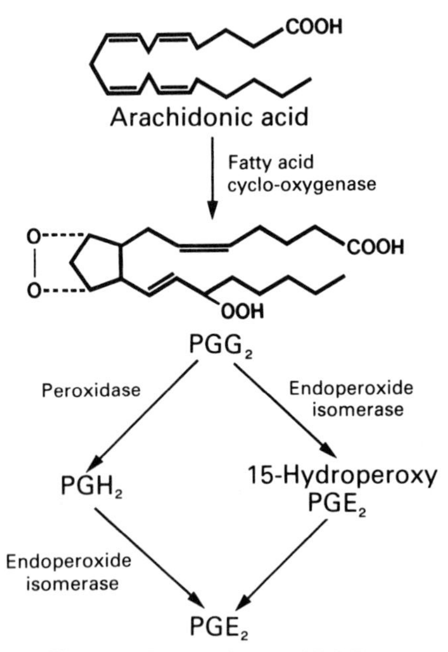

Fig. 3-2. Biosynthesis of PGE_2.

A_2 (with a half-life of 30–40 seconds), and prostacyclin (PGI_2). These are potentially highly potent blood-vessel and smooth-muscle regulatory agents, which are metabolized to the more stable and less active compounds, thromboxane B_2 and 6-oxo-$PGF_{1\alpha}$, respectively (Fig. 3-3). Prostaglandin E_2 and $PGF_{2\alpha}$ are themselves metabolized by a 15-hydroxy PG dehydrogenase (PGDH) to form C-15 ketone derivatives, which in turn are acted on by a Δ^{13}-reductase that reduces the double bond. Some of these prostaglandin metabolites may be biologically active. In addition, steroids, especially progesterone, are known to influence PGDH activity and thus the levels of primary prostaglandin.

An understanding of these pathways is important for evaluating studies in which inhibitors of prostaglandin synthesis have been used. Fig. 3-3 shows that whilst prostacyclin and thromboxane production from the cyclic endoperoxides can be

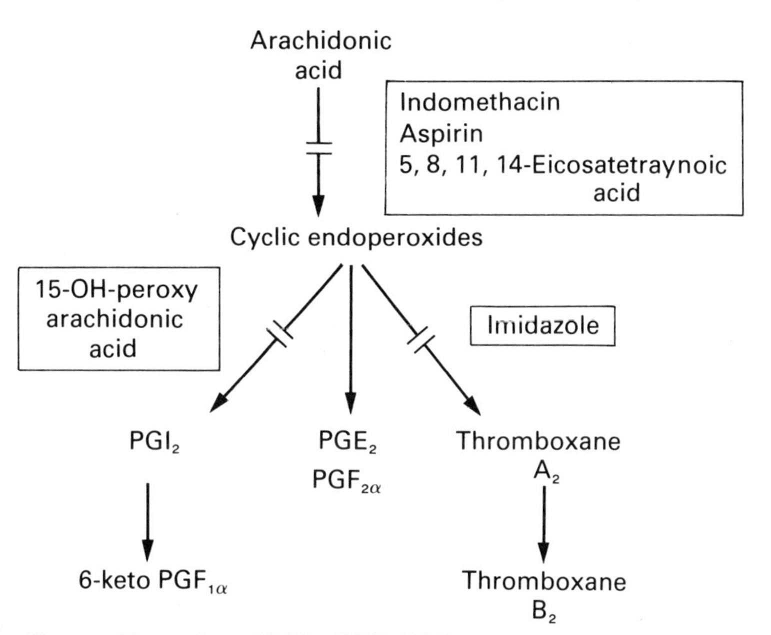

Fig. 3-3. Formation of PGI2, PGE, PGF and thromboxane A2 from arachidonic acid. Inhibitors of different enzyme systems are shown in the boxes.

blocked with specific inhibitors, generation of endoperoxide from arachidonate can be blocked with either substrate analogues such as 5, 8, 11, 14-eicosatetraynoic acid or with 'PG synthetase' inhibitors such as aspirin or indomethacin. There is now a vast literature on the use of these agents. However, because aspirin and indomethacin inhibit endoperoxide formation, they block prostacyclin and thromboxane production, as well as preventing prostaglandin generation. In order to interpret studies with such inhibitors, one must therefore take account of the potential importance of products of arachidonic acid metabolism other than PGE or PGF.

The use of inhibitors, the chemical synthesis of high-purity prostaglandins, and the development of sensitive and specific assay procedures for blood and tissues – all these have stimulated a considerable amount of research on prostaglandin action. In this chapter, we will discuss the role of prostaglandins in reproductive processes, but it should be remembered that these compounds are active in many parts of the body and may influence many other body systems, which in turn can exert effects on the reproductive system.

PROSTAGLANDINS IN THE MALE

Seminal vesicles are a major site of prostaglandin biosynthesis, and human seminal plasma contains a wide range of different derivatives, but particularly PGE_1 and PGE_2. Recently, the 19-hydroxylated derivatives of PGE_1 and PGE_2 have been demonstrated to be the chief forms in human semen. Production of prostaglandins in males is highly androgen-dependent. PG synthetase activity in the testis decreases after hypophysectomy and, in men, the prostaglandin concentration in the ejaculate falls within 48 hours after withdrawal of the testosterone treatment given to hypogonadal individuals.

High levels of prostaglandins in the male seem to be involved in some way with ejaculation and sperm transport. In the bull, rabbit, stallion and boar, injections of prostaglandins increase sperm number in the ejaculate and duration of ejaculation. In men, aspirin administration decreases the PGE and PGF content of semen, and it has been claimed by Swedish investigators that the concentrations of PGE in semen are significantly lower in cases of otherwise unexplained infertility. However, there is no correlation between prostaglandin concentrations in the ejaculate and sperm motility or oxygen consumption, or in fructose and lactate metabolism in spermatozoa. Furthermore, the smooth-muscle contractions associated with ejaculation do not depend on high concentrations of prostaglandins, since animals with little prostaglandin in the semen are still able to ejaculate, as are human subjects who have taken aspirin and so reduced their seminal PGE content by up to 80 per cent. The 19-hydroxy PGEs in human seminal plasma in fact depress the respiration of human spermatozoa.

Prostaglandins are present in the capsule of the testis, and exogenous $PGF_{2\alpha}$ produces contractions of the rabbit testis. However, there is still a paucity of information concerning the role of prostaglandins in spermatogenesis and sperm transport in the male. Of greater importance may be the role of seminal prostaglandins in sperm transport in the female reproductive tract. Physiological amounts of $PGF_{2\alpha}$, but not PGE_2, increase the penetration and motility of spermatozoa through human cervical mucus *in vitro*. However, no difference could be found in the prostaglandin concentration in seminal plasma of animals that ejaculate directly into the uterus, as compared to those that ejaculate into the vagina. In rabbits and sheep, exogenous prostaglandins hasten the transport of spermatozoa up to the oviduct, but the effect is presumably directly on the uterine musculature or ciliary activity, and from what we know at present this may not be of great physiological significance.

OVULATION AND GONADOTROPHIN RELEASE

The possible importance of prostaglandins in the process of ovulation (Book 1, Chapter 2) was first suggested from independent studies by Hal Behrman's group in the USA and Hans Lindner's group in Israel. They showed that treatment of rats with indomethacin could block the process of ovulation, and because chronic administration of indomethacin also blocked the induction of ovulation by LH, it was suggested that the major site of prostaglandin action was the Graafian follicle. We now believe that prostaglandins can also influence gonadotrophin release from the pituitary.

The follicular site of action was clearly implicated in rats when the indomethacin block of ovulation was shown to be overcome by follicular injection of PGE_2, but not of LH. Subsequently, investigators found that rats injected with the anaesthetic nembutal on the afternoon of pro-oestrus (thereby delaying the LH surge and ovulation by 24 hours) could be induced to ovulate by PGE_2, which also induced completion of meiosis in most

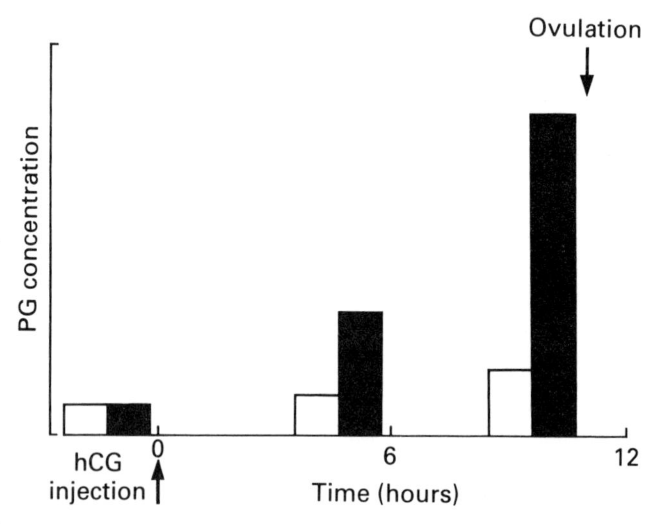

Fig. 3-4. Changes in follicular prostaglandin (PGE, PGF) concentrations after hCG injection in rabbits. Open histograms, animals pretreated with indomethacin to block prostaglandin synthesis. Solid histograms, control animals. In animals treated with indomethacin 30 minutes before hCG, the rise in prostaglandin concentrations and ovulation does not occur. (Modified from Le Maire *et al. Prostaglandins* **3**, 367–376 (1973).)

oocytes (a process described in Book 1, Chapter 2). These studies prompted the suggestion that prostaglandin action is normally exerted on the developing follicle, and that postaglandin production is stimulated by LH. Further support for this idea has been obtained in studies with the rabbit, a species in which ovulation occurs about 9–11 hours after human chorionic gonadotrophin (hCG) administration (Fig. 3-4). After hCG injection, the concentrations of PGF and PGE are elevated within 5 hours in follicles destined to ovulate, and a further increase is seen at 9 hours after injection. Treatment of animals with indomethacin before hCG injection blocked the increase and prevented ovulation. Moreover, microinjection of antibodies to PGE or PGF into the follicles also prevented ovulation. More recently, a similar increase in follicular prostaglandin concentration associated with ovulation has been described in the pig, and further studies from David Armstrong's laboratory in Canada have shown that after treatment with hCG the concentration of PGF is raised in human follicles maintained in organ culture. In the animal studies, prevention of prostaglandin synthesis by systemic injection of indomethacin results in the egg being trapped within a developing corpus luteum which is otherwise histologically normal and secretes progesterone.

These studies all point to the involvement of prostaglandins in mediating the induction of ovulation by LH in several animal species. The effect of LH in raising prostaglandin concentrations in the follicle is blocked by the protein synthesis inhibitor puromycin, suggesting that LH has to induce an increase in protein synthesis to bring about this effect.

At present, the actual mechanism by which prostaglandins might cause follicle rupture remains speculative – perhaps they induce the closure of blood vessels associated with stigma formation in the follicle wall. The demonstration of myosin fibres in the follicle wall by immunofluorescence is of interest since they would provide a target site for prostaglandin action. A further influence could be on the activity of hydrolytic enzymes implicated in follicular rupture.

Anna-Ritta Fuchs in New York has proposed yet another mechanism. She and her colleagues implanted recording balloons in the ovarian stroma of rabbits and measured the changes in intra-ovarian pressure during ovulations induced by mating or by injections of hCG. They found that ovarian contractility increased 5–7 hours after hCG treatment, which correlates well with the increase in $PGF_{2\alpha}$ content reported by David Armstrong's group. At the time of expected ovulation, the frequency of contraction increased greatly, and was sustained for several hours. The suggestion is therefore made that such contractions result in a squeezing action at the periphery of the ovary where the follicles are located; this could contribute to follicle rupture at the weakest point, even though follicular pressure may be unchanged (the mechanism of ovulation is discussed in Book 1, Chapter 2).

Prostaglandins may also influence gonadotrophin release. Injection of PGE_2 and $PGF_{2\alpha}$ into sheep, hamsters and rats causes an increase in the concentration of LH in peripheral blood, and as we have already seen, PGE_2 is able to overcome the block on LH release brought about in rats by nembutal administration. However, these studies do not indicate whether the principal site of this prostaglandin action is at the pituitary

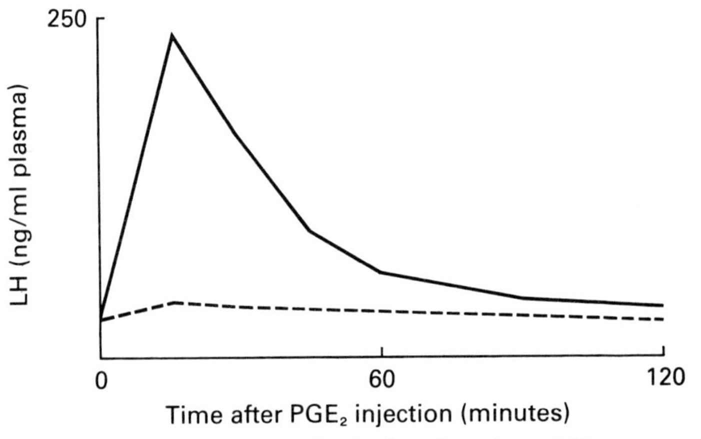

Fig. 3-5. Effect of PGE_2 injection (at time 0) on plasma LH concentration in control female rats (solid line) or rats treated previously with an antiserum to LH–RH (dashed line). (Modified from Labrie *et al.* In *Subcellular Mechanisms in Reproductive Neuroendocrinology*. Ed. F. Naftolin, K. J. Ryan and I. J. Davies, p. 391–406. Amsterdam; Elsevier (1976).)

or the hypothalamic level. The effect on LH levels produced by the simultaneous administration of PGE_2 and antibodies to luteinizing hormone releasing hormone (LH-RH) has been studied in the hope of resolving this question. Fernand Labrie and his colleagues in Canada have clearly demonstrated in anaesthetized rats that the administration of LH-RH antiserum not only reduced the basal LH concentrations, but also completely obliterated the rise in LH normally seen in response to PGE_1 or PGE_2 injection (Fig. 3-5). Their observations therefore suggest that the prostaglandins influence LH levels by an effect on LH-RH release. However, PGE increases cyclic AMP levels in both the median eminence and the pituitary (see Book 3, Chapter 2), and a hypophysial site of action cannot be entirely excluded.

This question has been examined further in experiments in which PGE_2 was injected directly into the third ventricle of the brain. The procedure increased the hypophysial portal and peripheral plasma LH-RH titres as well as LH release. Work from McCann's group in Texas has suggested that PGE_2 acts directly on the LH-RH secreting neurones to stimulate LH-RH release. He has demonstrated that the release of LH in response to PGE_2 is unaltered by blockers of the α-receptor, dopamine receptor, and serotonin and cholinergic receptors. When indomethacin was added to pituitaries incubated *in vitro* there was no effect on either basal or LH-RH-stimulated release of LH, which is also consistent with the hypothesis that the hypothalamus is the major site of prostaglandin action. When different prostaglandins were compared for their effectiveness in inducing LH release, it was found that whilst PGE_2 and $PGF_{2\alpha}$ were potent stimulators, PGE_1, $PGF_{1\alpha}$, PGA_2 and PGB_2 were relatively inactive. Comparison of their structures (Fig. 3-1) indicates the importance of a double bond in the 5, 6-position and an 11-hydroxyl group for activation of a brain receptor.

These studies also show that PGE is not working through conversion to other prostaglandins. McCann and his colleagues have explored the possibility that alternative intermediates in the biosynthetic pathway might also be effective in eliciting gonadotrophin release. They showed, however, that after intraventricular injection of analogues of the unstable endoperoxides there was little change in plasma LH or FSH. Thus, PGE_2 itself appears to be the active principle. Finally, we should note that in several species E and F prostaglandins also stimulate the release of growth hormone, prolactin, and ACTH. The actions of these other pituitary hormones may be of indirect importance to the reproductive processes.

GAMETE TRANSPORT AND TUBAL CONTRACTILITY

Prostaglandins are powerful stimulants of smooth muscle. We will consider later their role in uterine contractility and the events surrounding menstruation and birth, but for the moment we will focus attention on their actions on the uterus and Fallopian tubes, which might influence egg and sperm transport and implantation. Any mechanism that, for example, alters the rate of transport of eggs along the Fallopian tube, could be of potential importance as a contraceptive.

Early studies in sheep, human beings and rabbits led to the general conclusion that E prostaglandins relaxed, whilst F prostaglandins contracted, the Fallopian tubes, and these activities were shown to influence the rate of egg transport. Thus, in the rat, treatment with PGE delayed tubal transport, so that blastocysts could still be recovered from the oviducts 7 days after ovulation in treated animals. A peculiarity of this effect is the variability between species; in the rabbit, PGE_1 appeared to accelerate egg transport, although greater acceleration was found after administration of $PGF_{2\alpha}$. The response to exogenous prostaglandin is influenced by the circulating steroid concentrations. In ovariectomized rabbits, oestradiol treatment increased the amplitude and frequency of spontaneous contractions of oviductal muscle taken from the ampullary (ovarian) end of the oviduct. However, with muscle taken from the isthmic (uterine) region, progesterone, rather than oestradiol, increased the response to PGE_1. These differences probably relate to the important observation that different parts of the oviducts differ in their prostaglandin content. Thus, investigators have shown that $PGF_{2\alpha}$ predominates in extracts of human isthmic muscle, whereas PGE is the main prostaglandin in the ampullary region. This differential distribution could explain the mutually antagonistic effects of different prostaglandins. Thus, F prostaglandins may cause retention of eggs in the oviduct by tubal occlusion, whereas the E prostaglandins would allow passage of the eggs by tubal relaxation. The finding of specific binding sites for prostaglandins in rabbit oviductal tissue, and the demonstration that uptake is dependent on the hormonal state of the animal, have substantiated these early observations.

We referred previously to the recent interesting finding that the major prostaglandins of primate semen are 19-hydroxy derivatives. In tests in the pregnant rhesus monkey, 19-hydroxy PGF was found to have no effect on uterine contractility whereas, in strong contrast, 19-hydroxy PGE caused an increase in uterine tone and in the frequency of contractions. However, 19-hydroxy PGE produces great relaxation of human uterine muscle strips *in vitro*, and of rabbit uterine muscle *in vivo*. We do not understand the reasons for these interesting inter-species variations. The actions of prostaglandins are not confined to mammals, and recent experiments suggest that they perform similar functions in birds. In the domestic hen, for example, prostaglandins are clearly involved in egg laying (oviposition). PGE_1 or PGE_2 appears to exert a stimulatory effect on the uterus causing transport of the large egg to the utero-vaginal junction and the vagina, which both relax under the influence of PGE_1 in advance of the egg, thereby facilitating its passage.

IMPLANTATION

Prostaglandins may be involved in implantation. The early evidence was derived from studies in which administration of the PG synthetase inhibitor, indomethacin was shown to prevent or delay implantation. The effect of prostaglandins on the decidual reaction (see Book 2, Chapter 1) has recently been examined. Decidualization in the rat is characterized by the transformation of the stromal cells of the uterus into binucleated decidual cells, and this requires precise hormonal conditions and appropriate stimulation of the endometrium. Many substances induce decidualization in rats primed with progesterone and oestrogen but are ineffective in animals primed only with progesterone. However, scratching the endometrium is one way to induce the decidual response in such progesterone-primed animals. Workers in Etienne Baulieu's laboratory in Paris showed that the decidualization induced by scratching the uterus could be blocked by indomethacin administration. Furthermore, $PGF_{2\alpha}$ instilled in the uterine lumen of the prepubertal rat maintained on progesterone, induced a massive decidual response, the uterine weight of treated animals increasing by 350 per cent. The effect was not mediated through release of ovarian oestrogen, and could be reproduced in a dose-dependent fashion by intrauterine instillation of the precursor arachidonic acid. Recently 6-keto- $PGF_{1\alpha}$, a stable product of PGI_2, has been identified as the major prostaglandin present in the uterine tissues of several animals. Tom Kennedy and Jiri Zamecnik in Canada have shown that in the rat the concentration of 6-keto-$PGF_{1\alpha}$ is particularly elevated at those sites in the uterus where implantation will occur (see Book 2, Chapter 1). It is possible that PGI_2 influences the increase in vascularity at these sites, and that its inhibitory effect on smooth muscle is important in regulating uterine contractility.

GONADOTROPHINS AND STEROIDOGENESIS

The mechanism of gonadotrophin action on granulosa cells and on luteal cells is discussed in detail in Chapter 2 of the present volume and in Book 3, Chapter 1. It is of considerable interest that prostaglandins mimic many of the effects of gonadotrophins on isolated follicles and on granulosa cells. Thus, administration of PGE_1 and PGE_2 induces luteinization of isolated granulosa cells; PGE_2 will stimulate progesterone secretion by porcine granulosa cells, and both LH and the E prostaglandins stimulate an increase in cyclic AMP production by follicles or granulosa cells in culture. A key question that has aroused considerable controversy is whether prostaglandins are essential intermediates in the mechanism by which LH induces luteinization and steroidogenesis (Fig. 3-6*a*), or whether they only mimic the actions of LH. If this were the case, there should be separate membrane receptors for PGE_2 and for LH, as envisaged in Fig.

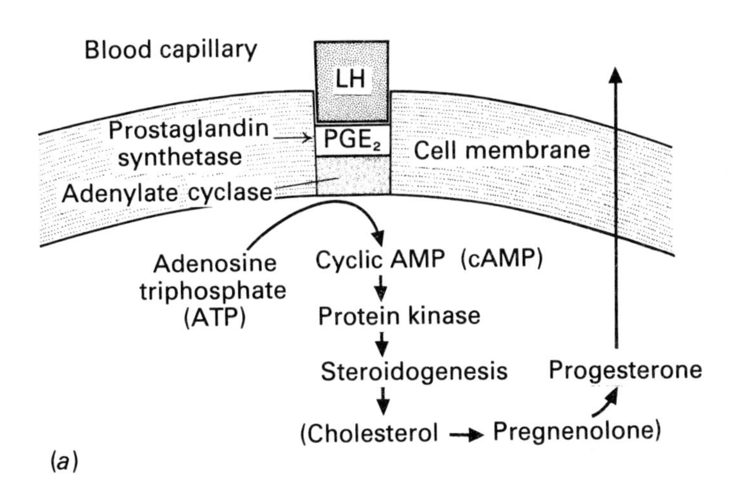

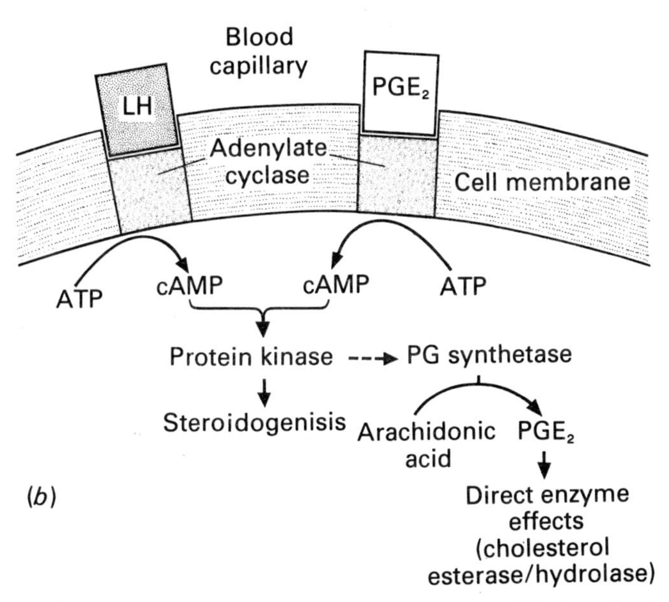

Fig. 3-6. Possible interactions of LH and PGE_3 in stimulating steroidogenesis. (*a*) LH acts via PGE_2. (*b*) LH and PGE_2 act independently to produce similar effects.

3-6(*b*). An additional possibility is that the LH-mediated increase in protein kinase (see Chapter 2) is itself responsible for the rise in PG synthetase activity (Fig. 3-6*b*). A hypothetical scheme such as that shown in Fig. 3-6*b* would explain why the increase in cyclic AMP is more rapid after LH administration than the increase in PGE_2. It would also explain why inhibitors of PG synthetase fail to block LH-mediated increases in cyclic AMP and steroidogenesis.

The concept that E prostaglandins may provide an essential link between trophic hormone binding and adenylate cyclase activation has been extended to the action of FSH on the testis, TSH on the thyroid, and ACTH on the adrenal. The key aspects of the interaction between LH, prostaglandins and steroidogenesis have been critically examined in an elegant series of studies from Hans Lindner's laboratory in Israel. His group has argued that if prostaglandins are essential intermediates of LH action, then: (i) inhibitors of PG synthetase activity should block the effects of LH on cyclic AMP generation and steroid production, and (ii) maximally effective amounts of LH and PGE_2 should not have additive effects on cyclic AMP levels. However, investigators in Lindner's laboratory found that although the rat ovary contained 'PG synthetase' activity, inhibition of this activity did not influence the stimulatory effect of LH on cyclic AMP production. In addition, LH had an effect on cyclic AMP levels within 20 minutes, whereas the same amount of gonadotrophin had no effect on PG synthetase activity, even after 1 hour. Furthermore the combined effects of maximally stimulating amounts of PGE_2 and LH on cyclic AMP accumulation were significantly greater than the effect of either compound alone, implying independent effects of LH and PGE_2 on cyclic AMP. Further evidence of separate sites of action was found in studies with ovaries from newborn rats. This tissue does not respond to LH with an increase in cyclic AMP, whereas it will respond to PGE_2; however, this difference in response possibly reflects the capabilities of different cell types within the various structures that make up the whole ovary. In summary, whilst prostaglandins can mimic many of the actions of LH on granulosa cells (including morphological transformation to luteinized cells, cyclic AMP generation and steroid production), this is probably via a discrete receptor mechanism, and not an essential component of LH action. As we shall see later, the different elements of the ovary contain both PG synthetase activity and specific prostaglandin receptors.

PROSTAGLANDINS AND CORPUS LUTEUM FUNCTION

Non-primates

The control of corpus luteum function and the importance of prostaglandins in corpus luteum regression was discussed by Roger Short in Book 3, Chapter 3. As he remarked, the non-pregnant sheep uterus is a major site of $PGF_{2\alpha}$ production. Thus, corpus luteum function is prolonged by hysterectomy, by autotransplantation of the ovary or the uterus to the neck, and by other surgical procedures that reduce the extent of vascular continuity between the utero-ovarian vein and the closely apposed ovarian artery. The local nature of this prostaglandin influence on the ovary has been demonstrated by performing a unilateral hysterectomy. The corpus luteum in the ovary that is contralateral to the remaining uterine horn is maintained, but the corpus luteum adjacent to the uterine horn regresses. Immunization of sheep or guinea pigs against $PGF_{2\alpha}$ prolongs the lifespan of the corpus luteum, as does intrauterine application of the PG-synthetase inhibitor, indomethacin. An increase in the concentration of PGF in the utero-ovarian vein is found at around the time of corpus luteum regression. In sheep with a 16–17-day oestrous cycle, significant PGF release occurs in a pulsatile fashion beginning around day 12–13. However, the major out-pouring of PGF into the utero-ovarian vein is only seen at the time of the final decrease in progesterone secretion from the regressing corpus luteum (Fig. 3-7). Similarly, a major release of PGF occurs in association with progesterone withdrawal after corpus luteum regression has been induced in the

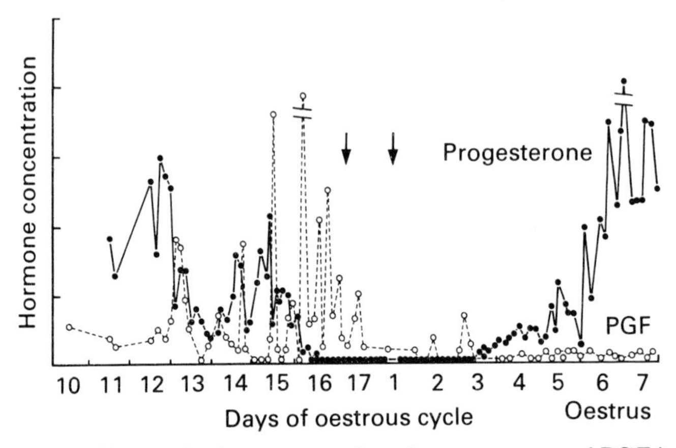

Fig. 3-7. Changes in the concentration of progesterone and PGF in the utero-ovarian venous blood of a sheep during the late luteal phase of the oestrous cycle. Note the pulsatile release of PGF, and the major release of PGF after regression of the corpus luteum. The arrows indicate the time of behavioural oestrus. (Modified from Thorburn *et al. J. Reprod. Fert.* Suppl. **18**, 151–158 (1973).)

middle of the oestrous cycle by injection of a suitable amount of a synthetic analogue of $PGF_{2\alpha}$ which is also luteolytic. Small but significant pulses of PGF are seen in utero-ovarian venous blood before day 12, and these seem likely to exert a tonic inhibitory effect on the corpus luteum, their action perhaps being balanced by circulating amounts of pituitary luteotrophic hormones.

There is ample evidence that $PGF_{2\alpha}$ causes regression of the corpus luteum in many non-primate species, and certainly in all of those in which a relationship between the uterus and luteal function has been demonstrated. Uterine $PGF_{2\alpha}$ production depends upon a period of progesterone priming, the progesterone apparently being necessary for activating the PG synthetase enzyme system. Ovariectomized sheep treated with progesterone alone have increased concentrations of PGF in their uterine tissues. However, if such animals also receive an injection of oestradiol, a further increase in prostaglandin production occurs. Oestradiol administration to ovariectomized animals that had not been pretreated with progesterone failed to stimulate any increase in uterine prostaglandin production. The role of progesterone in prostaglandin production, however, is paradoxical: whilst it is necessary for PG synthetase activity, and will cause an increase in PGF production, high levels of progesterone prevent PGF release, and withdrawal of progesterone allows PGF release to occur. The importance of oestradiol for uterine prostaglandin production was demonstrated in studies in which it was shown that if progesterone-primed sheep were immunized against oestradiol, or had been previously hysterectomized, then oestradiol failed to elicit an increase in PGF release. Extensive studies by David Baird in Scotland and John McCracken in the USA have indicated that the PGF secreted into the utero-ovarian vein passes by a counter-current mechanism through the walls of the vein into the ovarian artery. In the sheep, the artery is a tortuous vessel which closely adheres to the surface of the uterine vein.

Recent attention has focused on the mechanism by which $PGF_{2\alpha}$ induces luteolysis, many of the most illuminating studies being performed in the rat by Hal Behrman's group in the USA. In the rat, circulating LH is well known to bind to membrane receptors on the corpus luteum and thus stimulate progesterone synthesis. Using pseudopregnant rats, Behrman showed that $PGF_{2\alpha}$ induced luteolysis by directly antagonizing the action of LH, and by reducing the number of LH receptors. A rapid fall in the plasma progesterone concentration (70 per cent within 2 hours) was shown to occur after administration of $PGF_{2\alpha}$. Recent studies *in vivo* have demonstrated that the prostaglandin may act very quickly. Within 30 minutes after administration, $PGF_{2\alpha}$ starts to block the uptake of radioactively labelled hCG (which binds to the same receptor as LH) by the corpora lutea of pseudopregnant rats. Earlier studies had shown that, 24 hours after $PGF_{2\alpha}$ injection, the LH-binding capacity of the corpora lutea had dropped by 70 per cent, although the affinity constant for LH binding to the remaining receptors was unchanged.

The original suggestion was that $PGF_{2\alpha}$ exerted its luteolytic action by reducing ovarian blood flow, since it is a potent vasoconstrictor. However, most workers have now shown that the initial fall in progesterone precedes any change in total ovarian blood flow, although the possibility still remains of a change in the local distribution of blood flow within the ovary. In sheep and rabbits it was found, with the use of radioactive microspheres, that blood flow to the corpora lutea is reduced after $PGF_{2\alpha}$ administration whilst flow to the ovarian stroma and follicles increases, thus resulting in no net change in total flow.

Behrman's results make it likely that the early effect of $PGF_{2\alpha}$ is to restrict the ability of LH to increase cyclic AMP levels in the luteal cell. This action correlates well with the depletion of labelled hCG uptake observed *in vivo* within 20 minutes. It is important to recognize that $PGF_{2\alpha}$ does not affect basal cyclic AMP levels, but its action is to reduce the capacity of LH to stimulate any further increase in cyclic AMP. Because there appears to be an inverse relationship between intracellular cyclic AMP and cyclic GMP concentrations in a number of systems, the possibility was examined that ovarian cyclic GMP levels might be elevated after $PGF_{2\alpha}$. However, both LH and $PGF_{2\alpha}$ decreased the cyclic GMP concentration in Behrman's system, suggesting that the lowered cyclic AMP concentration was not due to production of the alternative nucleotide.

Within 2 hours after $PGF_{2\alpha}$ administration, a 30 per cent reduction in LH binding capacity was clearly demonstrable, both *in vivo* and *in vitro*. Antagonism of the trophic action of LH is therefore certainly a major effect of $PGF_{2\alpha}$. In the rat, $PGF_{2\alpha}$ also antagonizes the action of prolactin, which is of importance since exogenous prolactin alone will maintain luteal function in the hypophysectomized rat. In intact animals, the role of prolactin is apparently to maintain the LH receptor population. Administration of ergocryptine, a drug that suppresses prolactin, produces a similar decrease in LH receptors to that found after $PGF_{2\alpha}$ administration. This suggests that the prolactin–$PGF_{2\alpha}$ interaction at the ovarian level may be of major importance.

There is no evidence for a decrease in pituitary LH secretion after $PGF_{2\alpha}$ administration.

The results just described are consistent in part with the ideas put forward by Ken McNatty and Keith Henderson in Edinburgh. They suggested that $PGF_{2\alpha}$ suppressed cyclic AMP generation, although, unlike Behrman, they speculated that the prostaglandin effect was mediated through its own distinct receptor and was exerted in the membrane at a coupling site between LH binding and cyclic AMP generation. They argued that normally a major result of cyclic AMP generation and protein kinase activation is phosphorylation of the enzyme cholesteryl esterase, which converts cholesterol to pregnenolone. By blocking cyclic AMP, $PGF_{2\alpha}$ causes dephosphorylation of cholesteryl esterase and a reversion to the inactive form of the enzyme. Thus subsequent phosphorylation steps are slowed and the rate of steroidogenesis falls. Newly formed corpora lutea which are resistant to $PGF_{2\alpha}$ contain substantial amounts of bound LH and may be protected from the prostaglandin. Using tissue culture the Edinburgh group showed that higher amounts of $PGF_{2\alpha}$ were required to suppress progesterone production in luteinized granulosa cells stimulated with LH and FSH for 6 days previously, than in cells that had not been exposed to these gonadotrophins. The main difference between this model and that proposed by Behrman relates to the initial site of prostaglandin action: namely, whether the prostaglandin effect on cyclic AMP is mediated through an LH or a prostaglandin receptor mechanism. Because the membrane is a mosaic, it may transpire that both schools are right to a certain extent.

Luteolytic mechanisms probably vary between different species, particularly since the luteotrophic complex is not the same in each. $PGF_{2\alpha}$ treatment clearly influences the activity of one or more ovarian enzyme systems, but whether this action is independent of the early changes in LH binding and subsequent events including protein kinase activation is uncertain. In the rat corpus luteum, $PGF_{2\alpha}$ increases the activity of 20α-hydroxy

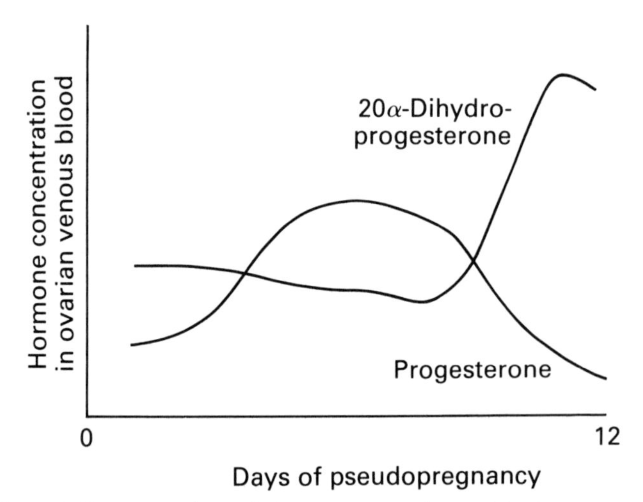

Fig. 3-8. Concentrations of progesterone and 20α-dihydroprogesterone in ovarian venous blood of the rat during pseudopregnancy. (Modified from Hashimoto *et al. Endocrinology* **82**, 333–341 (1965).)

steroid dehydrogenase. This enzyme is responsible for converting progesterone to a biologically less active metabolite, 20α-dihydroprogesterone; thus an increase in activity of this enzyme effectively shunts progesterone through an alternative pathway without necessitating an overall decrease in steroidogenesis. Activation of this enzyme produces an increase in the plasma concentration of 20α-dihydroprogesterone, as is seen at the end of pregnancy or pseudopregnancy in the rat, coincident with a decrease in progesterone concentration (Fig. 3-8).

Finally, we infer that the level of luteal function, as measured by the concentration of progesterone in peripheral blood, is really the result of an equilibrium between the action of positive (luteotrophic) and negative (luteolytic) agents. Recently, PGE_2, which is known to have opposite effects to $PGF_{2\alpha}$ in many systems, has been shown to prolong the lifespan of the ovine corpus luteum. Ovarian tissue has the capacity to synthesize prostaglandins, and eventually the balance between the different prostaglandins may prove to be the mechanism whereby the luteal cell regulates its progesterone secretion.

Primates

We still do not understand the mechanism for bringing about regression of the primate corpus luteum. In women and monkeys, normal ovarian cycles continue after removal of the uterus, thus raising the possibility that control of the corpus luteum is through a local intra-ovarian mechanism. Whether this involves prostaglandins is still somewhat speculative, for exogenous derivatives infused into women during the mid-luteal phase of the menstrual cycle produce only a transient fall in plasma progesterone. In women and monkeys, however, oestradiol is luteolytic, and rhesus monkeys that have received subcutaneous implants of oestradiol from the day after ovulation have significantly shorter luteal phases than control animals. The luteolytic effect of oestrogen in the monkey can be blocked with the PG synthetase inhibitor, indomethacin, implying that its action is mediated through prostaglandin production. Different groups have clearly shown that both the corpus luteum and the interstitial tissue in the monkey and human ovary contain PGF, and have the capacity to synthesize PGE and PGF from arachidonic acid. Membrane fractions from the human corpus luteum also have specific receptor sites for $PGF_{2\alpha}$, suggesting that this prostaglandin does have a physiological role in corpus luteum regulation.

At present, this fragmentary evidence is little more than suggestive of a role for prostaglandins in regression of the primate corpus luteum. However, an exciting new development in early 1977 was the finding by John Wilks that a metabolite of $PGF_{2\alpha}$, namely 15-oxo $PGF_{2\alpha}$, produced a rapid and sustained depression in the plasma progesterone concentration of non-pregnant rhesus monkeys. Although the amounts used were high (500 mg per monkey), results such as these raise the possibility that, in primates, a product of arachidonic acid metabolism other than $PGF_{2\alpha}$ might be responsible for lyteolysis. This would also explain the observation that there was no consistent trend in the concentration of PGF, nor in the ratio of PGE to PGF in corpora lutea taken at different stages of the human menstrual cycle. The possibility of an intra-ovarian mechanism in primates becomes more fascinating when we remember that the primate corpus luteum differs from the corpus luteum in many other animals in having the capacity to secrete oestrogen. In the monkey, the concentration of oestrone in the corpus luteum is highest during the late luteal phase and this oestrogen may turn out to be of critical importance, either in stimulating prostaglandin synthesis or in influencing the concentration of prostaglandin receptors in the corpus luteum membrane. Previous studies in human subjects treated with exogenous $PGF_{2\alpha}$ might have failed to produce luteolysis because inadequate amounts were administered. This could be of special importance because the corpus luteum also secretes PGE which is luteotrophic, and this effect may have to be overcome before luteal regression occurs. Because low levels of LH are required throughout the lifespan of the primate corpus luteum, an interaction of luteal prostaglandin with the LH receptor or with the adenylate cyclase system could produce luteolysis in a manner similar to that postulated for non-primates. Indeed McNatty and Henderson have shown that whilst progesterone secretion by cultured human granulosa cells can be stimulated by LH and FSH, addition of physiological amounts of $PGF_{2\alpha}$ suppresses steroid production to control levels.

Menstruation

As long ago as 1957, Pickles showed that extracts of human menstrual fluid possessed smooth muscle stimulatory activity. Prostaglandins are now known to be present in human endometrium obtained at any stage of the menstrual cycle. However, the concentration of $PGF_{2\alpha}$ is higher during the luteal than during the proliferative phase of the cycle, whilst the concentration of PGE reaches a maximum at the time of menstruation (Fig. 3-9). The high levels of prostaglandins in the endometrium coincide with the time of maximum contractile activity of the

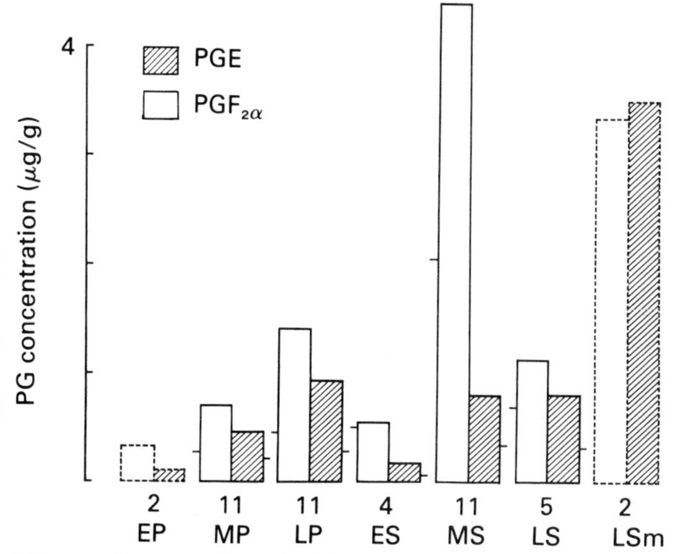

Fig. 3-9. Concentrations of $PGF_{2\alpha}$ and PGE_2 expressed as μg/g of fresh endometrial tissue obtained at different times in the menstrual cycle. EP, MP, LP: early, middle and late proliferative stages. ES, MS, LS: early, middle and late secretory stages. LSm, late secretory-menstrual stage. Figures along base line are the numbers of observations. (From Maathuis and Kelly. *J. Endocrinol.* **77**, 61–71 (1978).)

myometrium, and with the constriction of the spiral arterioles in the endometrium. Both these processes could be influenced by prostaglandins or by other products of arachidonic acid metabolism. The changing prostaglandin levels are probably under steroidal control, and the increase in uterine production of the hormone during the late luteal phase of the menstrual cycle may be related to the decline in the concentration of progesterone in blood. Oestradiol may also stimulate endometrial prostaglandin production, and after treatment of non-pregnant monkeys with oestrogen an increase in the concentration of PGF in uterine fluid has been found. In primates, uterine prostaglandins are apparently not involved in luteolysis, because normal ovarian cyclical activity continues after hysterectomy, and women with congenital absence of the uterus also have regular ovarian cycles (see page 103).

Many women experience primary dysmenorrhoea (painful menses). Several investigators have suggested that this condition, which is associated with excessive uterine muscular activity, may result from an imbalance between E and F prostaglandins, or excessive prostaglandin production. Administration of aspirin, or more potent PG synthetase inhibitors, relieves many of the symptoms, and provides convincing evidence that a product of arachidonic acid metabolism, which may be a primary prostaglandin, is involved in the aetiology of this condition.

PROSTAGLANDINS DURING PREGNANCY

The placentae of many animals both synthesize and metabolize prostaglandins. We can therefore reasonably ask whether these substances have any influence on placental endocrine function, and on uterine or placental blood flow.

There is only a limited amount of information on the effects of prostaglandins on placental steroid production. Prostaglandins do not appear to affect placental progesterone production in the same way that they influence luteal progesterone secretion, but they do stimulate oestrogen formation by the human placenta, which may stem from their ability to increase the formation of cyclic AMP from ATP by placental tissue. This may be of importance in relation to the cascade of endocrine interactions that is associated with the process of birth (see below).

Evidence is accumulating that prostaglandins may affect both uterine and placental blood flow. The same derivative may have different effects on different vascular beds, and different ones may have antagonistic effects on the same vascular beds. In studies in pregnant dogs, indomethacin administration resulted in a 25 per cent fall in uterine blood flow; this was accompanied by a 90 per cent decrease in the concentration of PGE in uterine venous blood. When bradykinin, a substance known to stimulate endogenous prostaglandin release, was injected into pregnant

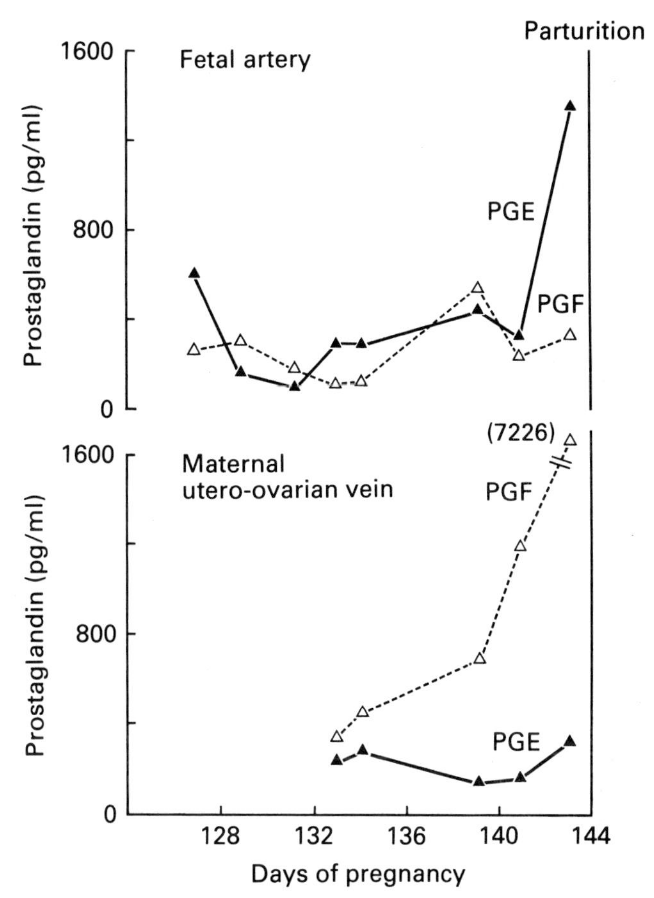

Fig. 3-10. The concentrations of PGE and PGF in the fetal femoral artery and the maternal utero-ovarian vein of the sheep during the last 20 days of pregnancy. Note the increase in prostaglandin concentrations at the time of parturition. The maternal utero-ovarian vein concentrations are much higher than they would be in the maternal peripheral plasma owing to the rapid rate of prostaglandin inactivation in the lung. (From Challis *et al. Prostaglandins* **11**, 1041–52 (1976).)

dogs, uterine blood flow was increased significantly, but this effect could again be abolished by concurrent indomethacin administration. In the sheep, studies from Rankin's laboratory in the USA have suggested that PGE_2 has a vasodilating action on the utero-placental circulation, and a vasoconstricting action on the umbilical–placental circulation.

During the last 3–4 years, the importance of prostaglandins in the regulation of fetal haemodynamics has been recognized; they have been measured in the femoral arterial blood plasma of fetal lambs *in utero*, and in the umbilical cord blood of human fetuses. Although there are considerable technical problems associated with measuring PGE and PGF in blood, most investigators have found that their concentrations in fetal plasma are higher than those in maternal peripheral plasma, collected under similar conditions (Fig. 3-10).

Prostaglandins in cord blood could result from production by the cord musculature itself. They may be of importance in regulating blood flow through the cord and in the rapid closure of the umbilical circulation that occurs at birth. PG endoperoxides and thromboxanes are even more active than $PGF_{2\alpha}$ or PGE_2 in causing constriction of the umbilical artery, and these compounds may be the biologically active molecules. An important function of prostaglandins during fetal life is to maintain the patency of the ductus arteriosus, the blood vessel that diverts blood past the fetal lung directly into the descending aorta. Administration of E prostaglandins dilates the ductus, particularly at low oxygen tensions. The relative importance of prostaglandins produced in the vessel itself or present in the fetal circulation still remains to be established. After birth the ductus normally closes, a process that has been attributed to the change in oxygen tension, but that may be mediated by a prostaglandin (such as PGI_2) produced locally by the muscle of the vessel wall. The relationship between prostaglandins and the ductus arteriosus has been exploited clinically. The infusion of PGE_1 has been used to dilate the ductus arteriosus in certain newborn infants with forms of cyanotic heart disease, in which it is desirable to maintain ductal patency between the time of diagnosis and corrective surgery.

PROSTAGLANDINS AND PARTURITION

If prostaglandins are involved in the normal physiological events associated with birth, then one might expect three requirements to be met: (i) their concentration should increase at the time of parturition, (ii) inhibitors of their synthesis should block uterine contractility and prolong pregnancy, and (iii) exogenous prostaglandin should stimulate uterine contractility. For a number of animal species, including man, all these conditions can be fulfilled.

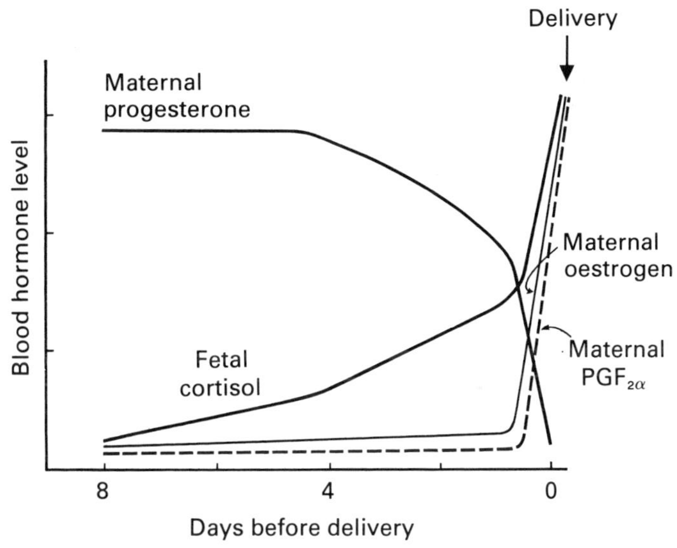

Fig. 3-11. Changes in the concentrations of cortisol in fetal blood and of progesterone, oestrogen and $PGF_{2\alpha}$ in maternal blood in relation to the time of parturition of sheep.

In the sheep, the time of birth is determined by the fetus. The pioneering studies of Mont Liggins have been reviewed by him in Book 2, Chapter 3 of this series and the endocrine events leading up to parturition in the sheep are summarized in Fig. 3-11. Briefly, activation of the fetal hypothalamo-pituitary, adrenal axis during the final 10–15 days of pregnancy is believed to result in an increase in cortisol secretion by the fetal adrenal gland. This cortisol acts on the placenta where it induces certain steroid-metabolizing enzymes, resulting in a decrease in placental progesterone production and an increase in placental oestrogen production. The changes in the levels of these steroids, particularly the increase in oestrogen production, stimulate $PHF_{2\alpha}$ synthesis and its release from the maternal portion of the placenta and from the myometrium. The increase in $PGF_{2\alpha}$ is believed to be responsible for triggering the final sequence of contractile events in the smooth-muscle cells of the myometrium. Recent studies have also shown changes in the production of $PGF_{2\alpha}$ synthesis and its release from the maternal portion of the However, the physiological importance of these changes in relation to those in $PGF_{2\alpha}$ concentrations remains to be established.

We now know that an increase in $PGF_{2\alpha}$ production can be brought about by a variety of different stimuli, both endocrine and mechanical. This may offer animals a backup or 'fail safe' mechanism should any one particular trigger fail. Briefly, prostaglandin production at term can be provoked by the same stimuli as are effective in non-pregnant animals. Thus, an increase in oestrogen or a decrease in progesterone can prove potent stimuli to prostaglandin release (Fig. 3-12). Oxytocin, released from the maternal posterior pituitary as a result of the neuroendocrine reflex initiated when the fetal head engages in the cervix and vagina, also causes an increase in prostaglandin production. Other factors, such as 'stretch' of the uterus, and rupture of the fetal membranes, probably contribute also to the raised PGF levels seen at term.

An increase in prostaglandin production at term is seen in sheep and primates. In man and monkeys, the main source of these compounds is probably the fetal membranes and decidua. The mechanism by which the sudden increase in their production is effected is not known, but appears to involve either an increase in the availability of arachidonic acid, or an increase in the enzyme activity responsible for converting arachidonic acid to prostaglandins.

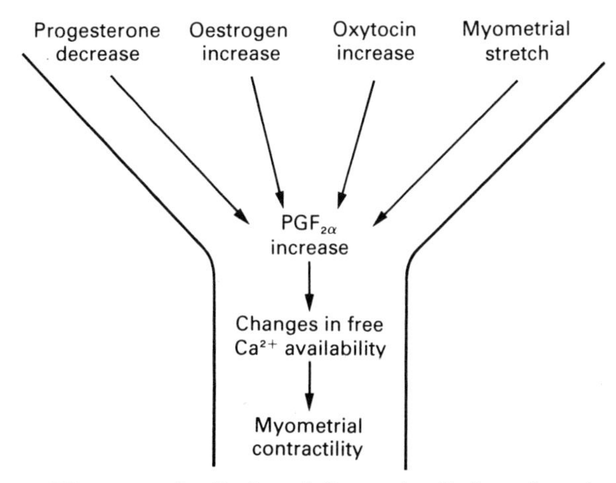

Fig. 3-12. The prostaglandin funnel. Prostaglandin is envisaged as a key step in the initiation of myometrial contractility. Its production can be increased by several different stimuli.

Studies with inhibitors of prostaglandin production have been used to provide evidence for a role of endogenously-produced derivatives in the birth process. In monkeys, administration of indomethacin prolongs pregnancy by up to 20 days. Clearly, one cannot do such an experiment in women, but in patients who suffer from rheumatoid arthritis, and who have taken aspirin through much of pregnancy, the length of gestation is prolonged, and the duration of labour itself is lengthened.

Prostaglandins have at least two functions associated with parturition. In species such as the rabbit and goat, in which progesterone from the corpora lutea is essential throughout gestation for the maintenance of pregnancy, prostaglandins have a luteolytic role. Elevated levels of $PGF_{2\alpha}$ are found in the utero-ovarian venous blood of the goat just before the normal regression of the corpus luteum at term. Administration of $PGF_{2\alpha}$ to pregnant goats results in cessation of corpus luteum function and in premature delivery. In many species, irrespective of the site of progesterone production, $PGF_{2\alpha}$ stimulates myometrial activity at term. The primate uterus is exquisitely sensitive to exogenous prostaglandins, and contracts in response to them even in early pregnancy when it will not respond to oxytocin.

The precise mechanism by which prostaglandins stimulate myometrial activity is not known. Work in Stan Korenman's laboratory in Los Angeles has established the sequence of biochemical changes that occur when smooth-muscle cells contract or relax. Korenman has suggested that agents that block uterine activity (β-agonists) increase intracellular cyclic AMP levels, resulting in changes in cyclic AMP-dependent protein kinase activity, which in turn may result in an increase in ATP-dependent Ca^{2+} transport by the cell membranes. Prostaglandins could work on either the cyclic nucleotide or calcium step in this mechanism. It has been found that the increase in intracellular cyclic AMP concentration, brought about by β-agonists such as isoproterenol, can be inhibited by PGE_2 and $PGF_{2\alpha}$, the latter being more potent. This activity correlates with the specific binding of prostaglandins to the myometrium in different animal species. It is clear that prostaglandins can influence the flow of Ca^{2+} in and out of the myometrial cell. Carsten in Los Angeles has shown that $PGF_{2\alpha}$ reduces the amount of Ca^{2+} that is bound to intracellular membranes. The concentration of free Ca^{2+} available for coupling–contraction with the myosin filaments is therefore increased, and contraction occurs. The intracellular Ca^{2+} concentration is also influenced by the intracellular cyclic nucleotide concentration, since flux of the ion across the cell membrane is an ATP-dependent process.

Recent work has indicated that at term prostaglandins not only make the uterus contract but may also induce the changes in the cervix that are necessary for delivery. Clinical trials are in progress to establish the value of prostaglandins in dilating the human cervix.

PROSTAGLANDINS AND ABORTION

By the fourth month of human pregnancy, the uterine lumen becomes fully occupied by the amniotic sac. The decidua capsularis, the tissue immediately outside the chorionic membrane, fuses with the decidua parietalis, the tissue lining the

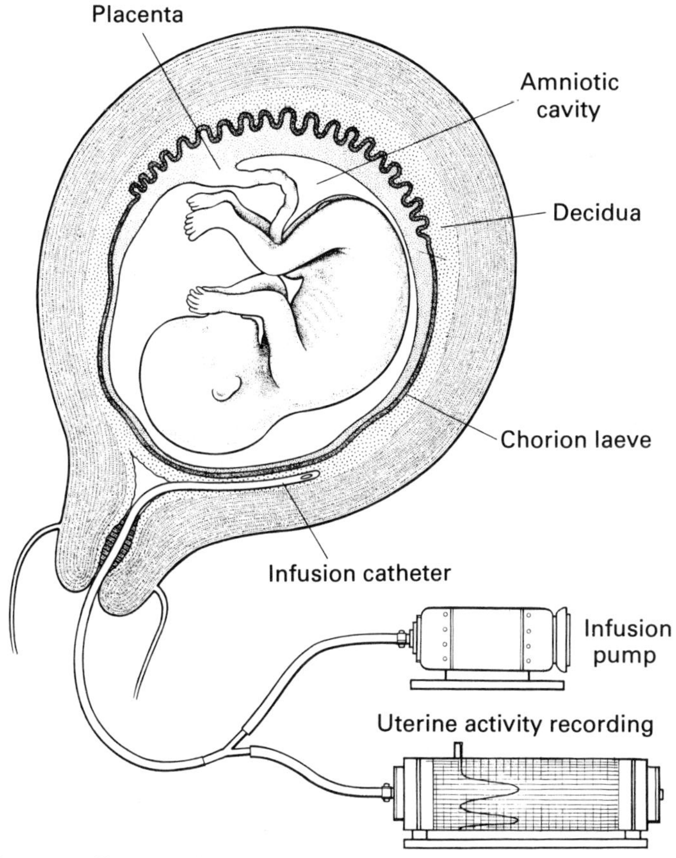

Fig. 3-13. Extra-ovular (extra-amniotic) catheter placement for inducing abortion by prostaglandin infusion. The agent penetrates between the layers of decidua, outside the amniotic cavity. Because uterine contractions may result in expulsion of the catheter, more recent methods have involved a single injection of PGE_2 or $PGF_{2\alpha}$ directly into the amniotic cavity.

uterine wall opposite to the site of placental attachment. The cells of the decidua, in contrast to cells derived from trophoblast tissue, are very fragile and susceptible to osmotic shock. Bjorn Gustavii from Sweden proposed in 1973 that after the intra-amniotic administration of hypertonic saline to induce abortion, the lysosomes of the decidual cells rupture, releasing the enzyme phospholipase A_2 and increasing the biosynthesis of prostaglandins. The prostaglandins labilize more lysosomes, and, more importantly, make the uterus contract, thereby inducing abortion.

Abortion can also be induced in mid-pregnancy by administration of prostaglandin. Originally the prostaglandin was infused outside the amniotic cavity between the decidual cell layers (Fig. 3-13). More recent methods have involved a single injection of PGE_2 or $PGF_{2\alpha}$ directly into the amniotic cavity. This method circumvents the problems of catheter displacement that were encountered with the extra-amniotic infusion. Numerous trials have now started with the intra-vaginal administration of prostaglandins in suitable gels. Not only does this afford the simplest and safest route of administration, but it may induce local cervical dilation whilst minimizing undesirable systemic side-effects.

PROSTAGLANDINS IN THE FUTURE

During the last decade, the discovery of the central role of prostaglandins, not only in reproductive processes but in so many other facets of physiology, has led to the rapid application of basic knowledge to clinical practice. Their luteolytic properties have been exploited for the development of new methods for synchronizing oestrus in horses, cattle and sheep and for inducing parturition in corpus-luteum-dependent species such as pigs. In the laboratory, work has concentrated on producing various analogues with longer half-lives and a reduced incidence of unwanted side-effects such as nausea and vomiting that were encountered in the early clinical trials of prostaglandins in human subjects. Prostaglandins are effective for inducing labour at term and mid-trimester abortion. In very early pregnancy they stimulate uterine contractility and impair the endocrine function of the conceptus. Indirectly therefore, they may cause regression of the human corpus luteum, during those critical 20 days after ovulation before the placenta makes enough progesterone to maintain pregnancy. Already trials have shown that prostaglandins self-administered vaginally can induce menstruation in either the non-pregnant or early pregnant patient. Such a procedure may totally eliminate the need for hospitalization for early abortion, and could be of enormous value in population control.

Our knowledge of the prostaglandins and their relevance in reproductive processes has made extraordinary progress in recent years, and further studies will surely continue to shed new light on many basic physiological mechanisms. The challenge that will face us to understand the information gained and translate it, whenever possible, into human and animal clinical practice, offers a truly exciting prospect.

SUGGESTED FURTHER READING

Gonadotrophin action on cultured Graafian follicles: induction of maturation division of the mammalian oocyte and differentiation of the luteal cell. H. R. Lindner, A. Tsafriri, M. E. Liberman, U. Zor, Y. Koch, S. Baumbinger and A. Barnea. *Recent Progress in Hormone Research* **30**, 79–138 (1974).

Prostaglandins in reproduction. B. B. Pharriss and J. E. Shaw. *Annual Review of Physiology* **36**, 391–412 (1974).

Role of prostaglandins in reproduction. V. J. Goldberg and P. W. Ramwell. *Physiological Reviews* **55**, 325–51 (1975).

Inhibition of uterine motility: the possible role of the prostaglandins and aspirin-like drugs. K. I. Williams and J. R. Vane. *Pharmacology and Theraputics* **1B**, 89–113 (1975).

Rapid block of gonadotrophin uptake by corpora lutea *in vivo* induced by prostaglandin $F_{2\alpha}$. H. R. Behrman and M. Hichens. *Prostaglandins* **12**, 83–95 (1976).

Prostaglandins

Uterine luteolytic hormones: a physiological role for prostaglandin $F_{2\alpha}$. E. W. Horton and N. L. Poyser. *Physiological Reviews* **56**, 595–651 (1976).

Endocrine control of parturition. G. D. Thorburn, J. R. G. Challis and J. S. Robinson. *Biology of the Uterus*, pp. 653–732. Ed. R. M. Wynn. New York; Plenum Press (1977).

4 The androgens

W. I. P. Mainwaring

The castration of male farm livestock has been a common practice since the beginning of recorded history. Examples of this are still seen today in the capon, ox and gelding, which are considered to have desirable attributes lacked by their intact counterparts, the cockerel, bull and stallion. Castration of the human male was also carried out for a variety of reasons, punitive, religious or even aesthetic. In Europe, young boys – the castrati – were treated before puberty in order to retain their treble voices in cathedral choirs, and in the Middle East, men were made eunuchs so that they could serve as harem keepers.

From the time of the ancients, studies on the capon have helped to lay the foundations of endocrinology. About 350 B.C. Aristotle recorded the regressive changes in the cockerel after castration, including loss of fertility, aggression, sexual behaviour and secondary sexual characteristics such as the comb and the spurs (Fig. 4-1). These observations indicated the wide physiological influence of the gonads. But it was not until 1849 that Berthold made what may be considered the first endocrine experiment by transplanting the testis from a cockerel into a capon; he noted how the stunted comb of the capon slowly began to grow again. In 1911, Pezard demonstrated that the injection of a saline extract of the testis was also effective in restoring comb growth in capons.

Great efforts were then made by chemists to isolate the male-promoting substances or androgens (from the Greek: *aner*, male and *genos*, descent), and in 1935, Leslie succeeded in purifying the steroid hormone, testosterone, from the bull's testis. Subsequent experiments have established beyond reasonable doubt that it is the primary androgen secreted by the testis of all higher animals.

Fig. 4-1. The influence of testosterone on the growth of the comb and wattles of the cockerel; capon on the left, intact rooster on the right.

THE ACTIONS OF ANDROGENS

One of the great fascinations of the androgens is that they evoke such a wide spectrum of changes in the male animal at all stages of development, not simply at the time of puberty. The traditional view that testosterone promotes the growth of certain organs still remains at the centre of current thinking, but any comprehensive theory of the mechanism of androgen action must also take into account the diverse effects listed in Table 4-1. This list is not intended to be comprehensive, nor is it invariably applicable or to all species. For example, the differentiation of the brain is extremely complex and only in a few rodents and primates does testosterone promote this process in the neonatal period of development; in other species, differentiation of the brain occurs before birth. In addition, the wide range of

BLE 4-1. The biological effects of androgens on target organs of male mals

Organ	Species	Developmental period	Effect
ernal genitalia g. penis)	All	Embryonic	Sexual differentiation
essory sex glands g. prostate)	All	Pubertal	Rapid growth and stimulation of secretions
tis	All	Pubertal	Spermatogenesis
in	Most	Fetal or neonatal	Sexual differentiation
in	All	Adult	Male libido
er	Most	Embryonic	Haemoglobin synthesis
stoderm	Birds only	Embryonic	Haemoglobin synthesis
er	Most	Neonatal	Synthesis of enzymes
ney	Mouse	Adult	Enzyme synthesis and cellular hypertrophy
vary gland	Pig	Adult	Pheromone production
scle	Most	Pubertal	Slow growth (anabolic effect)
r follicles in ecific areas	Most	Pubertal	Hair growth
aceous glands	Most	Pubertal	Sebaceous secretion
e marrow	Most	Adult	RNA and protein synthesis
al cords	Most	Pubertal	Thickening of cords

secondary sexual characteristics of the male has been the subject of extensive research, the features studied including differences in the growth of deer antlers, and the courtship behaviour patterns in birds. In the human male, some current findings on voice and hair are absorbing yet not fully understood. Bass and baritone singers have been clearly shown to have a higher concentration of testosterone in their blood (plus a lower oestrogen level) than their more highly-pitched tenor colleagues. On the other hand, despite a universal similarity in the concentration of blood testosterone in all races, hair growth is ethnically distinct, with Orientals having significantly less pubic and bodily hair than Caucasians. A fascinating contrast is seen in Japan: the Ainou of the northern islands are of different stock from their southern countrymen and are characteristically hirsute.

A fashionable trend during the last few years has been for biochemical studies on androgen action to be carried out on the accessory sexual glands, especially the prostate. But we should not forget that, as far as reproductive success and the survival of the species are concerned, the stimulation of spermatogenesis is the most important role of testosterone; the secretions of the accessory organs are relatively unimportant for fertility. Further, since muscle constitutes such a major proportion of the male's body weight, the anabolic effects of testosterone, although poorly understood, are of great significance.

The influence of a steroid hormone can be elucidated by two means. The more direct and classical one involves extirpation of the organ of synthesis, followed by injection of the hormone into the depleted animal. The majority of experiments on androgen action have been conducted by restoration of the androgenic milieu in castrated animals with testosterone. Ebo Nieschlag and collaborators adopted the alternative approach, namely active immunization of the male against testosterone. They injected intact male rats, rabbits and rhesus monkeys with testosterone conjugated to bovine serum albumin, thereby causing the animals to produce high titres of an antibody that bound free testosterone. As a result, the circulating levels of free testosterone fell, and the concentrations of total (i.e bound) testosterone rose enormously. That this bound fraction was biologically inert was demonstrated by increased secretion of follicle stimulating hormone (FSH) and luteinizing hormone (LH) from the pituitaries of the immunized animals, with consequent testicular hypertrophy and the maintenance of spermatogenesis. Thus immunization proved to be an invaluable means for probing the important feedback mechanisms involved in regulating the synthesis of steroids such as testosterone.

About 1960, the commercial availability of isotopically-labelled steroids and precursors of nucleic acids and proteins heralded a new era of research, and enabled investigators to probe the molecular mode of action of androgens. Two significant contributions to our present knowledge warrant particular mention in a historical context. First, Wells Farnsworth in Buffalo had the insight to recognize that testosterone was extensively metabolized in many of its target organs; in the prostate, the principal metabolite was 5α-dihydrotestosterone. Second, Jean Wilson in Dallas established from studies on the preen gland of the drake that the interaction between testosterone (and presumably its metabolites) with target cells involved specific androgen-binding proteins or receptors. Following in the wake of these two pioneers, many other investigators have contributed enormously over the last decade to our knowledge of the mechanism of action of androgens, particularly Shutsung Liao and Guy Williams-Ashman (Chicago), Vidar Hansson and Kjell Tveter (Oslo) and Nicholas Bruchovsky (Edmonton). From this body of information, a plausible model for the mode of action of androgens can be built around the selective metabolism and subsequent binding of different metabolites in different target cells. Some examples of this working hypothesis are presented in Fig. 4-2.

While superficially attractive, the hypothesis presented in Fig. 4-2 is something of an oversimplification in two important respects. First, the Leydig (interstitial) cells in the testis are not the only sources of testosterone and related steroids: the adrenal cortex produces considerable amounts of androstenedione which can be converted to testosterone in the peripheral circulation. For experimental purposes, castration dramatically curtails the supply of testosterone, but for certain clinical purposes, including treatment of carcinoma of the prostate, complete suppression of androgen synthesis is often necessary. One technique is to implant needles of radioactive ^{99}yttrium into the pituitary, thereby suppressing the release of FSH, LH and ACTH into the peripheral circulation, and so inhibiting the secretion of androgens by the testis and the adrenal cortex.

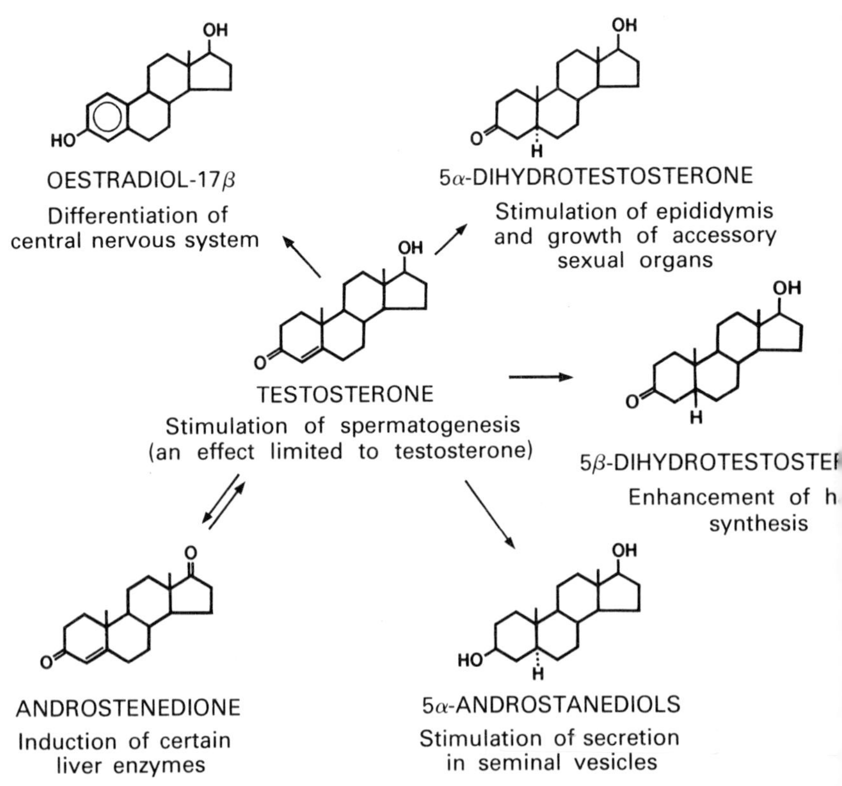

Fig. 4-2. The structure of testosterone and its biologically active metabolites, with an indication of their functions.

Testosterone itself can be considered as both a hormone and a prehormone. It is a hormone in the classical sense of Starling (1905) in that it acts in its own right, without metabolism, and it is a prehormone when it acts as the precursor of 5α-dihydrotestosterone or other biologically active metabolites. To assume that a given androgen target cell responds to only one particular metabolite of testosterone may be incorrect. In the rat prostate, for example, Etienne Baulieu and Ilse Lasnitzki have presented evidence that many metabolites are responsible for the overall androgenic response, each metabolite having a specified role of its own. In their view, 5α-androstanediols regulate secretion, 5α-dihydrotestosterone regulates cell division, and testosterone had as yet unspecified functions which may include regulation of the available receptor sites for the pituitary hormone, prolactin, on the plasma membrane.

BASIC MODEL FOR THE MECHANISM OF ANDROGEN ACTION

The model presented in Fig. 4-3 is derived primarily from studies on rat ventral prostate. Particular importance is attributed to the intracellular formation of 5α-dihydrotestosterone from testosterone, its binding to a specific cytoplasmic receptor protein, and the translocation of the 5α-dihydrotestosterone–receptor complex into the nucleus. Current evidence tends to suggest that testosterone does not play a particularly important part in the androgenic response of the rat prostate; in truth, its role in this target cell remains something of an enigma. Evidence from many laboratories suggests that testosterone has little, if any, affinity for the prostate androgen receptor, and unlike 5α-dihydrotestosterone, it cannot promote the attachment of the receptor protein to chromatin in reconstituted, cell-free systems. Nevertheless, testosterone can bind to chromatin directly *in vitro* in the complete absence of the cytoplasmic androgen receptor, yet the significance of this binding remains doubtful. Anti-androgens prevent the attachment of 5α-dihydrotestosterone to the receptor protein and thereby suppress virtually all androgen-mediated responses in rat prostate; this evidence tends to discredit the view that testosterone plays a direct or major part in the androgenic response in this particular accessory sexual gland.

Transport

Testosterone is firmly bound to proteins in the plasma, probably in all male animals. The Atlantic salmon and the thorny skate have plasma testosterone concentrations a hundred times higher

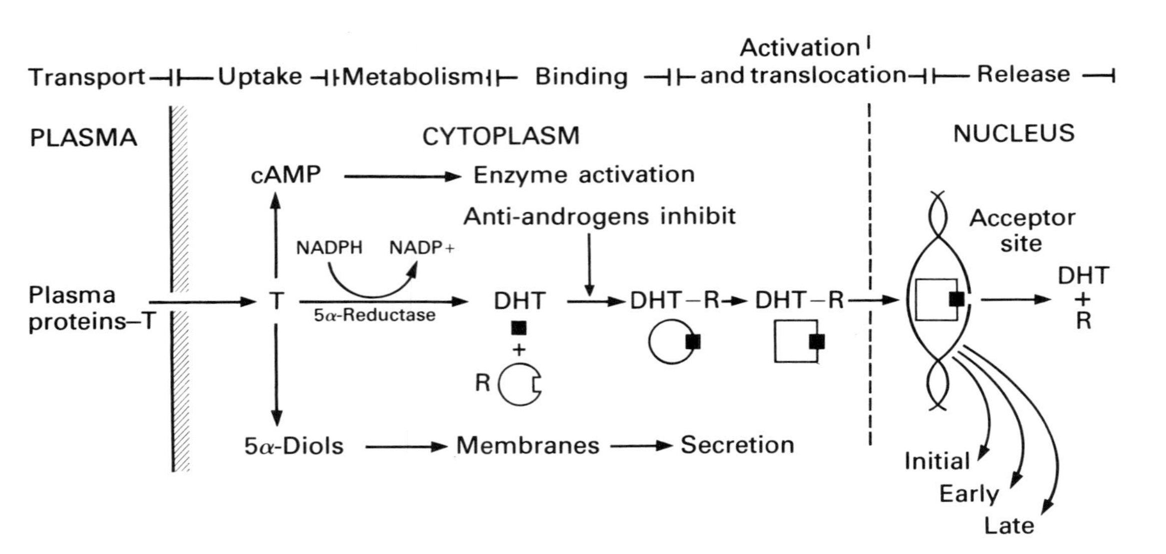

Fig. 4-3. The mechanism of action of androgens in rat prostate. The critical steps are the formation and nuclear binding of 5α-dihydrotestosterone (DHT). There is little evidence that testosterone plays a major part in evoking the androgenic response. Minor events are the synthesis of cyclic AMP, leading to the activation of certain enzymes, and a possible involvement of 5α-androstanediols in the secretory process. R, receptor protein; T, testosterone. The binding site in the receptor protein is indicated as the square niche, which is subsequently occupied by 5α-dihydrotestosterone, ■; a change in the configuration of the receptor complex is necessary for translocation to the nucleus, indicated by the change from ring to square.

than the human male, yet excessive androgenization is prevented by a special testosterone-binding protein. In some mammals, testosterone is bound to a sex steroid-binding β-globulin (SBG) in plasma, distinct from that in fishes. In those mammals, like the rat, that do possess SBG, testosterone is transported by other plasma proteins (Book 3, Chapter 1). Only the small proportion of free or unbound testosterone in plasma is biologically active; the remainder is biologically inactive, but in equilibrium with the free fraction. As described earlier, the experiments of Ebo Nieschlag with testosterone-specific antibodies illustrated the crucial biological importance of free plasma testosterone. This point has been emphasized further by studies on the effects of SBG on the response of rat prostate to testosterone in organ culture. In these experiments, Ilse Lasnitski found that by increasing the concentration of SBG in the culture medium, the effects of the exogenous androgen were suppressed; the prostate cells were finally deprived of androgens and regressed.

Uptake

Testosterone enters all cells of the body by passive diffusion, but there is some evidence that it may have a facilitated entry into androgen target cells by an energy-dependent or active transport system. Prolactin appears to assist in this uptake process.

Metabolism

The NADPH-dependent enzyme, 5α-reductase, located mainly in the endoplasmic reticulum, converts testosterone to 5α-dihydrotestosterone and some of this is converted further into 5α-androstanediols (for structures, see Fig. 4-2). The diols are strongly bound to the endoplasmic reticulum membranes and may play some part in controlling secretion in the prostate. Importantly, the diols remain in the cytoplasmic compartment and never reach the cell nucleus.

Binding

It is important to distinguish clearly between the systemic hormone testosterone, distributed in the plasma by binding to SBG or other transport proteins, and the subsequent binding of the hormone 5α-dihydrotestosterone, formed locally within the target cell itself. The 5α-dihydrotestosterone is selectively and very tightly bound to the androgen receptor protein, forming a discrete protein–steroid complex. The receptor protein has little, if any, affinity for non-metabolized testosterone; the binding ratio of 5α-dihydrotestosterone: testosterone is at least 10:1. In addition, the binding affinity of the receptor system is at least an order of magnitude higher than that of the plasma transport system. The specificity and higher efficiency of the receptor system means that the systemic hormone, testosterone, is almost 'sucked' into some androgen target cells, where the active metabolite, 5α-dihydrotestosterone, is bound selectively and with high affinity. As far as one can tell, there is only one androgen receptor protein in rat prostate, and this is specific for 5α-dihydrotestosterone rather than testosterone or the 5α-androstanediols. Testosterone continues to enter the target cells and be metabolized until the intracellular receptor sites are saturated with locally formed 5α-dihydrotestosterone. The receptor–dihydrotestosterone complex can be formed under artificial conditions at 4 °C *in vitro*, but at this low temperature it cannot interact with nuclear chromatin. At the temperature within mammalian cells, 37 °C, the receptor–dihydrotestosterone complex rapidly changes its physicochemical configuration and in some way becomes activated. Only in this activated state can the receptor complex be bound to nuclear chromatin, and this temperature-dependent change, imbuing the receptor complex with different properties, is critically important and we have simulated this in reconstituted, cell-free systems *in vitro*. The fundamental basis of the model is that non-target organs, such as spleen or lung, do not contain androgen receptor proteins. In these organs, testosterone may enter passively to some extent, but neither metabolism nor selective binding occurs.

Translocation

The activated complex then migrates from the cytoplasm through the pores of the nuclear envelope and occupies specific sites within nuclear chromatin called acceptor sites. Their nature is not absolutely clear, but they contain DNA and non-histone nuclear proteins. There are generally more acceptor sites in androgen target cells than non-target cells, and the nuclear binding of androgens, other than 5α-diols, is a basic feature of their mechanism of action. The translocation process is totally dependent on the presence of the receptor protein; it is a unique and characteristic property of this protein. Prostate chromatin cannot retain 5α-dihydrotestosterone in the absence of the receptor protein and all attempts to simulate the occupancy of the chromatin acceptor sites with SBG or other plasma proteins, in place of the receptor, have always failed. The plasma proteins fulfil an important transport function, but only the receptor protein can be considered as an important intracellular regulator.

Biochemical responses

The receptor complex triggers a series of biochemical events which are all fundamentally directed towards an increase in the transcription of the genetic information stored in the DNA of the chromosomes. The stimulation of transcription provides more ribosomal RNA, the essential machinery for protein synthesis; more messenger RNA, to specify the amino acid sequence of proteins made during the androgenic response; and more DNA, in preparation for cell division during the final phase of glandular growth. As indicated in Fig. 4-3, the responses elicited by the receptor complex occur at different rates and may be subdivided into initial, early and late events. Nevertheless, they are closely integrated, and components synthesized early on are needed for later processes. The initial production of ribosomal RNA is a prerequisite for protein synthesis and this in turn is essential for DNA synthesis, a late response.

Release

By definition, most hormone responses must be reversible or of finite rather than infinite duration. Accordingly, the receptor complex occupies the acceptor sites for up to 12 or 16 hours, and then leaves or is degraded by processes yet to be elucidated. The exit of the receptor complex makes room for newly-formed complex, provided that the supply of 5α-dihydrotestosterone and its receptor are maintained. The receptor protein itself is either synthesized continuously or recycled from nucleus to cytoplasm. Evidence supporting these alternatives has been put forward, but the precise nature of the release mechanism remains in some doubt. If the intracellular movement of steroid hormones is viewed as a series of dynamic and interrelated processes, incoming receptor complex would displace its forerunner from the nucleus by the law of mass action.

The model attaches crucial importance to the receptor mechanism. As yet, no one has succeeded in purifying the androgen-receptor protein and so its precise role in regulation remains to be established. None the less, considerable indirect evidence supports the model:

(*a*) Many anti-androgens are available, and these compete with 5α-dihydrotestosterone for the binding site(s) on the receptor protein and thus suppress the stimulation of events mediated by the receptor system. So few processes are insensitive to the blocking action of anti-androgens that the receptor system can safely be said to be of paramount importance.

(*b*) Studies on mutants with inherited defects give unique insights into control mechanisms operative in biology. The testicular feminization mutant (*Tfm*) mouse, first described by Mary Lyon and Susan Hawkes at Harwell, is important in the present context (see also Book 6, Chapter 1). This mutant has a normal male XY karyotype, yet has a female phenotype. A gene situated on the X chromosome results in the complete loss of the androgen receptor system from all the target cells and the animals are insensitive to even massive doses of 5α-dihydrotestosterone or testosterone. In the mouse, the submaxillary salivary gland is a useful indicator of androgen responsiveness, since it is sexually dimorphic and secretes male-specific proteins, including proteases, only when stimulated by testosterone. Castration of males leads to the disappearance of these salivary gland proteins, yet the male-specific proteases may be induced even in females after injections of testosterone. Mary Lyon found that the submaxillary gland of the *Tfm* mouse is completely refractory to testosterone; the male-specific proteases can never appear because of the deletion of the androgen receptor system. Similar findings apply to another mutant, the pseudohermaphrodite (*Ps*) rat. Mary Lyon and her co-workers have also used the *Tfm* mouse in a fascinating way to explore the influence of testosterone on spermatogenesis. With the sophisticated technique of embryo aggregation, chimaeras or 'artificial' mice can be produced by fusing early embryos with different genotypes. The investigators at Harwell obtained male mouse chimaeras by fusing androgen-resistant (*Tfm*/Y) and normal (XY) mouse embryos. When adult, the mice were tested for their fertility. These chimaeras produced spermatozoa from both the normal and the *Tfm* germ-cell lines, suggesting that although testosterone is essential for spermatogenesis, it has no direct action on the germ cells. Presumably it must act just on the Sertoli cells, which in turn regulate germ-cell activity.

(*c*) A number of experimental tumours are available, with sublines that are either sensitive or insensitive to androgens in cell culture. Without exception, the sensitive tumours contain androgen receptors, whereas the insensitive tumours do not.

(*d*) Finally, the specificity of the binding of steroids matches very closely the biological activity of the steroids in a given cell type. This is too frequent and precise to be considered fortuitous.

Despite the plausibility of the model, there are serious difficulties in applying it lock, stock and barrel to all other androgen target organs, even to the prostate of other species. As we shall see, variations in the metabolism of testosterone and the selective binding of certain metabolites can explain the differential nature of the androgenic responses in different cells.

ANDROGEN METABOLISM AND BINDING IN DIFFERENT TARGET CELLS

We can gain some idea of the importance of the structure of a steroid for its biological function by examining the synthetic steroids in Fig. 4-4. By carefully manipulating the structure of testosterone, we can produce more powerful androgens, anti-androgens or anabolic steroids. By definition, the latter have only a marginal influence on the growth of the accessory sex organs, such as the seminal vesicles, but selectively maintain the growth of skeletal muscle. The response of the bulbocavernosus (or less correctly, levator ani) muscle is widely used experimentally to monitor the anabolic activity of steroids, as it responds more quickly than most skeletal muscles. The most powerful anabolic steroid, stanazol, has a bulbocavernosus to seminal vesicle growth ratio of 30:1. The synthetic androgen, R1881, does not bind to plasma proteins, such as SBG, so it is therefore widely used in the specific assay of the androgen receptor proteins.

Similar arguments can be applied to the structure-function relationships of the various metabolites of testosterone produced in different target cells, especially oestradiol-17β, 5α-dihydrotestosterone and 5β-dihydrotestosterone. To emphasize their structural differences, their overall shapes are illustrated in Fig. 4-5 and may be compared directly with testosterone. Ring systems containing six carbon atoms, as in most of the structures in Fig. 4-5, are usually in what is called the chair form. This is more stable and less strained that the alternative boat form but

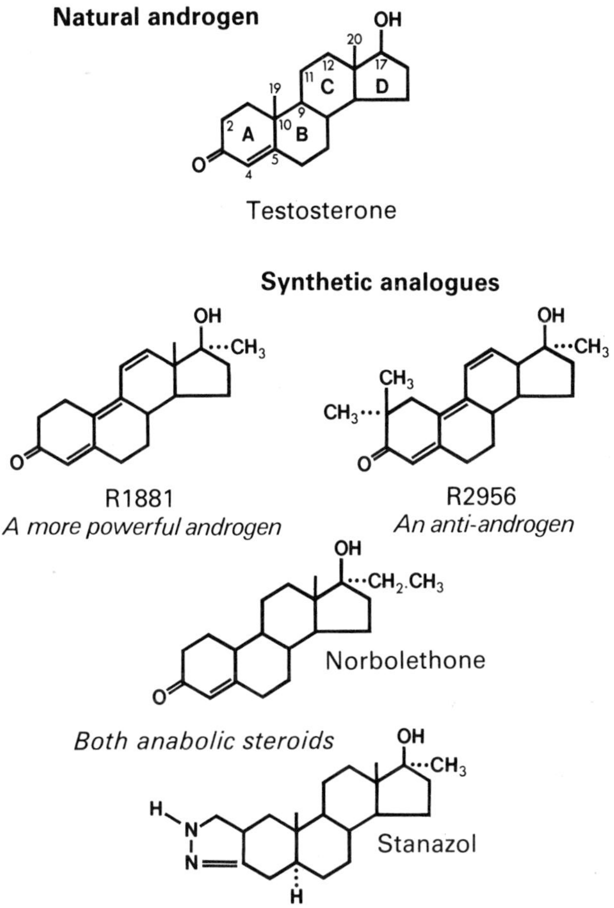

Natural androgen
OH
20
12
17
11
C
D
19
9
2
A
10
B
O
5
4
Testosterone
Synthetic analogues
OH
CH3
R1881
A more powerful androgen
OH
CH3
CH3
CH3
O
R2956
An anti-androgen
OH
CH2.CH3
Norbolethone
O
Both anabolic steroids
OH
CH3
H
N
N
H
Stanazol

Fig. 4-4. The influence of steroid structures on biological activity. R 1881 and R 2956 are code numbers of the Roussel Pharmaceutical Company. The numbers in the ring system of testosterone indicate the positions where structural modifications are important; dotted lines, bonds in α-orientation and solid lines, bonds in β-orientation.

the latter configuration is not found in naturally occurring steroids. In sharp contrast, the aromatic or benzene ring is perfectly flat, and five-membered rings, such as the steroid D ring, are always slightly distorted. On comparing these structures

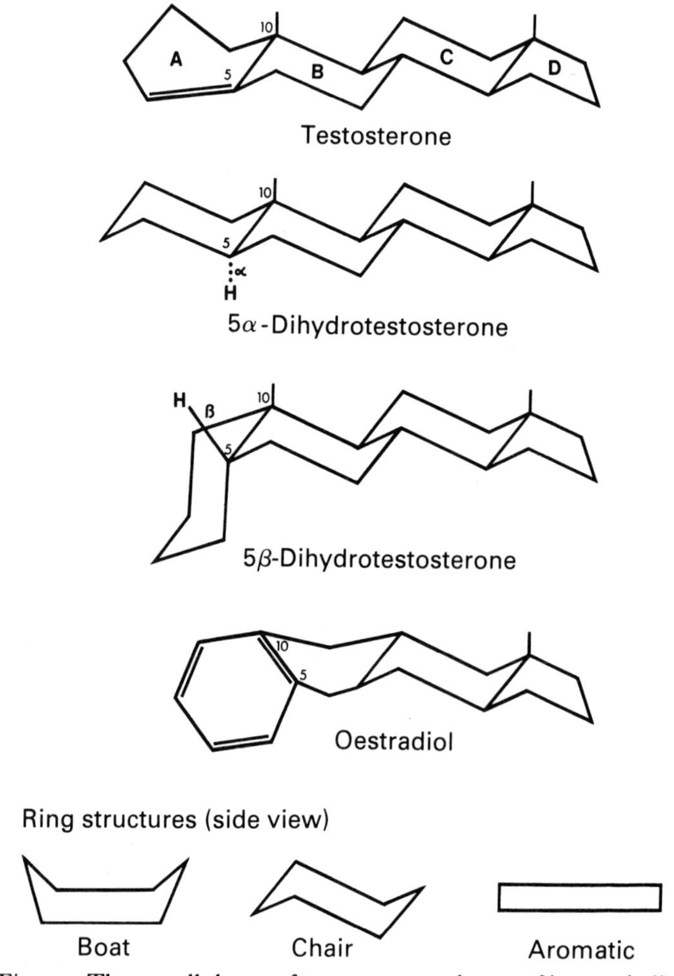

Fig. 4-5. The overall shapes of testosterone and some of its metabolites. For clarity, substituent groups are not included, but the C-5–C-10 axis is indicated.

carefully, the following points emerge. Testosterone itself is a relatively planar molecule, but the $\Delta^{4,5}$ double bond imposes strain and distortion in the A ring. The insertion of two hydrogen atoms into the double bond promotes a large change

TABLE 4-2. The formation and binding of testosterone metabolites in androgen target cells

Target tissue	Metabolite	Enzyme responsible	Binding		Main response	Comments
			Cytoplasm	Nucleus		
Embryonic anlagen	5αDHT	5α-Reductase	?	?	Differentiation	Anlagen specific for 5α-DHT or testosterone, not both androgens
Accessory sex organs	5α-DHT	5α-Reductase	+	+	Growth and secretion	5α-Reductase can show marked developmental changes e.g. bull prostate
Dog prostate	5α-Androstane-3α, 17α-diol	5α-Reductase, dehydrogenases	+	+	Growth	Unique metabolite
Testis	5α-DHT	5α-Reductase	+	+	Growth	5α-Reductase active for a short time only during sexual maturation
Fetal and neonatal brain	5α-DHT and oestradiol-17β	5α-Reductase aromatase	+	+	Sexual differentiation	Enzymes particularly active in neonatal period
Fetal liver and chick blastoderm	5β-DHT	5β-Reductase	+	–	Haemoglobin synthesis	No nuclear binding mechanism
Neonatal liver	Androstenedione,	17β-Dehydrogenase	+	+	Imprinting of enzymes	Androstenedione is not bound in adult liver
Muscle	None	No metabolism	+	–	Anabolic effect	No nuclear binding mechanism
Bone marrow	None	No metabolism	–	+	RNA synthesis	Failure to detect cytoplasmic receptor (technological problem?)
Hair follicles and sebaceous glands	5α-DHT, oestradiol-17β	5α-Reductase Aromatase	+	+	Growth of sex hair and secretions	Complex interaction, many receptors
Kidney	5α-androstanediols	5α-Reductase, dehydrogenases	+	+	Induction of enzymes	Complex mechanism, not always involving receptors
Comb, wattles	5α-DHT	5α-Reductase	+	?	Growth	Historically important

5α-DHT, 5α-dihydrotestosterone; ?, receptors present but location not proven; + or –, presence or absence of receptors.

between the structures of 5α- and 5β-dihydrotestosterone. In the 5α-isomer, with a hydrogen atom in the *trans* (or same side) orientation to the C-10 methyl group across the C-5–C-10 axis, the molecule becomes even more planar and the A ring is restored almost to the chair form. In the 5β-isomer, with a hydrogen atom in the *cis* (or opposite side) orientation to the C-10 methyl group, the molecule is directed into an extremely angular shape. In oestradiol-17β, the flat aromatic A ring imposes a major distortion or puckering of the adjacent B ring, which together with the loss of the angular methyl group at C-19, produces a molecule of totally different shape to testosterone. It is extremely difficult to draw the structure of oestradiol in a flat (two-dimensional) manner, but the perfectly planar A ring in oestradiol can be visualized by comparing the side views of chair, boat and aromatic rings at the foot of Fig. 4-5.

The importance of the different shapes of the possible metabolites of testosterone can now be grasped. The receptor concept demands a close geometric fit or structural recognition between the binding site on the receptor protein and its optimum steroid ligand, as with a key and lock. If androgen target cells have the ability to produce differently shaped metabolites of testosterone and also have distinctive receptors for recognizing these steroids, then this could explain how testosterone has such a diversity of biological effects. Present evidence, summarized in Table 4-2, essentially supports the concept of distinctive metabolic and binding properties in a wide range of androgen target cells.

Three aspects of Table 4-2 should be emphasized. First, each metabolite is implicated in one specific type of response. Second, the receptor mechanisms can be surprisingly tissue-specific. The dog prostate, for example, appears to bind 5α-androstanediol preferentially rather than 5α-dihydrotestosterone; this is a unique example of an androstanediol reaching the nucleus of a target cell. Furthermore, in both blastoderm and muscle, there is evidence favouring a cytoplasmic binding mechanism but without subsequent translocation to the nucleus. Third, certain biological responses of a more complex nature, such as the characteristic plumage of the two sexes in birds, are not included. While involving androgens, these responses also require systemic changes in the concentration of oestrogens.

None of the receptors listed in Table 4-2 has been purified. They all have similar characteristics, being labile, present in only minute amounts and displaying extreme preference for their favoured ligand. Their regulatory functions are not precisely known and their presence can only be inferred from the binding of radioactive steroids. However, their physicochemical properties are similar. Without exception, they are relatively large molecular weight 1.0–2.0×10^6), acid (pH 4.6–6.0), asymmetric (frictional ratio, f/f_0, 1.40–1.96) and bind the appropriate ligand with a very high affinity (dissociation constant, K_D 0.5–1.0×10^{-9}–10^{-10} M) but limited capacity.

ANTI-ANDROGENS

The pharmaceutical industry has produced several anti-androgens for the treatment of a variety of clinical disorders ranging from hypersexuality, acne, or prostatic cancer in men to hirsutism in women. The structures of two common anti-androgens are presented in Fig. 4-6. Many anti-androgens, such as cyproterone acetate, have the basic steroid skeleton, but others such as flutamide, are not steroids. Non-steroidal anti-androgens are somewhat flexible, with sufficient rotation round certain bonds to enable them to fit the binding site on the androgen receptor molecule. The resulting receptor–anti-androgen complex may even be translocated into the nucleus, but unlike the complex with 5α-dihydrotestosterone, it is not biologically active.

The influence of the structure of anti-androgens on their biological activity can be surprisingly subtle, a good example being cyproterone acetate, where the presence of the 17α-ester group profoundly modifies the properties of the steroid ring. Cyproterone acetate has progestational activity, which enables it to have an inhibitory feedback effect on the hypothalamus, as

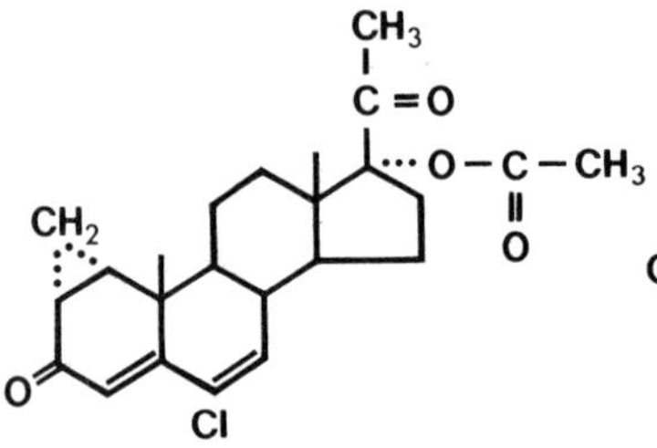

Cyprotene acetate

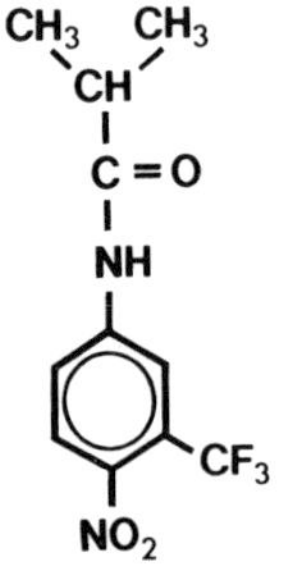

Flutamide

Fig. 4-6. The structure of two anti-androgens. Most anti-androgens inhibit reactions dependent on the formation of 5α-dihydrotestosterone and those requiring unmetabolized testosterone; selective antagonists for reactions that are dependent on the formation of other metabolites have yet to be developed.

well as giving it the properties of an androgen antagonist. By contrast, cyproterone itself has no progestational activity and promotes hypothalamic stimulation by inhibiting the negative feedback inhibition of testosterone. Cyproterone is much less active as an anti-androgen than its acetate.

No perfectly satisfactory anti-androgen has been developed to date and probably none is sufficiently specific for long-term use. All anti-androgens tested thus far have undesirable (and inevitable?) side-effects in men, such as loss of libido and breast development (gynaecomastia). These effects are perhaps not surprising, because most anti-androgens are indiscriminate in inhibiting all the responses controlled by 5α-dihydrotestosterone and testosterone. In fact, the effects of cyproterone acetate are so general that it is worth listing the few processes that are refractory to it: (*a*) the effects of testosterone in the uterus and vagina; (*b*) tubular hypertrophy and the synthesis of β-glucuronidase in the kidney of the mouse, processes that may depend on the binding of 5α-androstanediols to the cytoplasmic membranes; (*c*) the neonatal programming of certain organs, described later; (*d*) the activation of certain enzymes promoted by the intracellular formation of the classical 'second messenger', cyclic AMP. This cyclic nucleotide is inseparable from the mode of action of polypeptide hormones but it has far less significance in terms of steroid hormone action.

As the importance of the metabolism of testosterone has become more apparent, the rather surprising effects of certain steroid antagonists can now be explained. The synthetic compound, MER-25, is primarily an anti-oestrogen, yet it can also prevent the masculinization of the neonatal rat brain by testosterone. This process requires the conversion of testosterone to oestradiol-17β within the hypothalamus and the binding of the oestrogen to oestrogen receptors. It is this binding of oestradiol in the hypothalamus that is selectively prevented by MER-25.

BIOLOGICAL RESPONSES TO ANDROGENS

The responses described here have been selected to underline the importance of the formation and binding of different metabolites of testosterone.

Responses in the embryo

(*a*) *Development of sex organs.* As described in detail elsewhere (Book 2, Chapter 2; Book 6; Chapter 1) the primary sex organs, the testis and ovary, arise from a common pool of uncommitted cells in the embryo. In a manner not clear, the Y chromosome dictates that these cells develop into an embryonic testis, and in its absence an embryonic ovary forms. Some of the male and female accessory sex organs arise from primordial structures (or anlagen) derived from primitive Müllerian and Wolffian ducts, whereas the male and female external genitalia originate from common anlagen situated outside the primitive ducts (Fig. 4-7). The secretion of testosterone from the embryonic testis, with

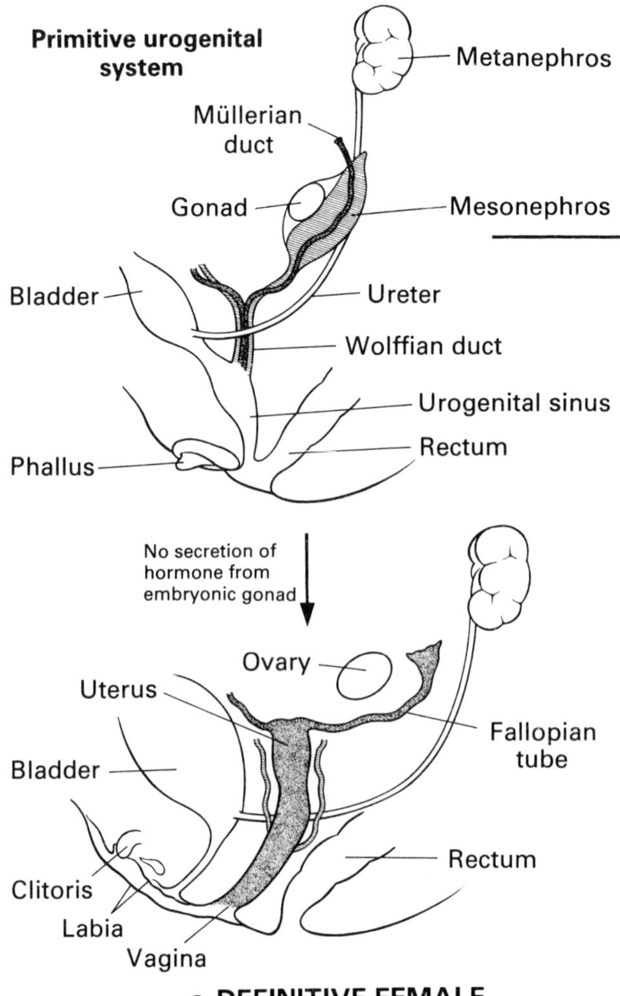

Fig. 4-7. Differentiation of the urogenital tract in males and females. In the process of differentiation to the male (*b*), structures requiring testosterone for differentiation are in italics, those requiring 5α-dihydrotestosterone are boxed.

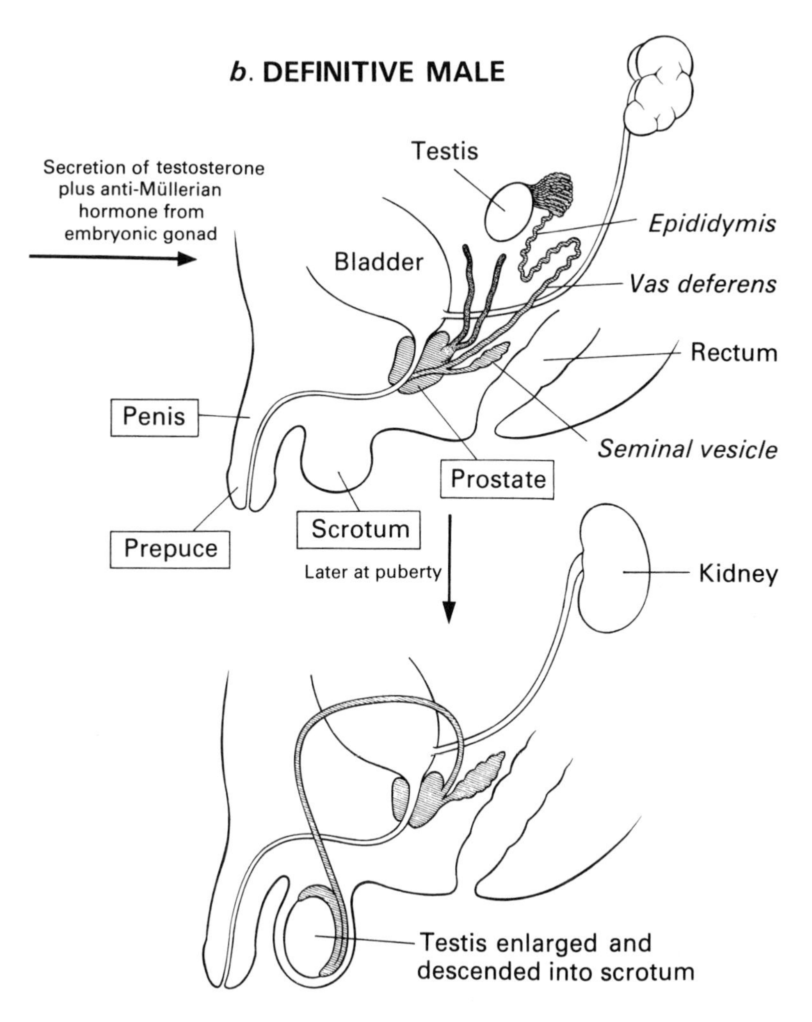
b. DEFINITIVE MALE
Secretion of testosterone plus anti-Müllerian hormone from embryonic gonad
Testis
Epididymis
Bladder
Vas deferens
Rectum
Penis
Seminal vesicle
Prostate
Scrotum
Prepuce
Later at puberty
Kidney
Testis enlarged and descended into scrotum
c. SEXUALLY MATURE MALE

Fig. 4-7 (cont.)

subsequent metabolism and binding of the metabolites, is responsible for the development of the Wolffian duct into the male reproductive tract, and the Müllerian duct regresses under the influence of a polypeptide hormone produced by the Sertoli cells. The Wolffian duct has no 5α-reductase activity and the binding of testosterone itself promotes the appearance of the embryonic epididymis, vas deferens and seminal vesicle. The remaining anlagen have an active 5α-reductase and the selective binding of 5α-dihydrotestosterone directs the embryonic appearance of the prostate and male external genitalia. Without the hormonal stimulation of the embryonic testis, the Wolffian duct atrophies completely, leaving the Müllerian duct to develop into the Fallopian tubes, uterus and cervix; the vagina and external female genitalia then develop from the common anlagen that in the male become the prostate and male external genitalia. In *Tfm* individuals, where the androgen receptors are absent, the embryonic testis secretes testosterone normally, but since none of the cells can respond to it, sexual differentiation is impaired and the individual becomes a sterile intersex, with testes, no uterus, and a female external phenotype.

There is another fascinating clinical condition in man, recently discovered in a village in the Dominican Republic; this is a type of intersexuality caused by greatly impaired activity of 5α-reductase. Such individuals have normal blood concentrations of testosterone and consequently they have essentially a male phenotype. But since all processes dependent on 5α-dihydrotestosterone are impaired, the external genitalia at birth are hypoplastic and the penis may be ill-formed (hypospadias); in later life, growth of the prostate and the appearance of the male pattern of facial and bodily hair may also fail to occur. At puberty, the increased secretion of testosterone stimulates growth of the penis and development of the skeletal masculature, and the onset of spermatogenesis.

(*b*) *Fetal haemoglobin synthesis.* In many birds, including the domestic fowl, the synthesis of haemoglobin begins in the fertilized egg in the delicate blastoderm, which surrounds the primitive streak. The blastoderm may be dissected out and maintained in culture, where it will eventually synthesize fetal haemoglobin. This process can be stimulated by 5β-reduced steroids, such as 5β-dihydrotestosterone, which have a unique and extremely angular structure. Our recent studies suggest that these steroids promote haemoglobin synthesis by a mechanism that does not involve the cell nucleus; the following findings support a cytoplasmic or translational means of control: (*a*) The blastoderm can actively metabolize testosterone to 5β-reduced steroids and only these angular steroids stimulate haemoglobin synthesis *in vitro*; the more planar 5α-reduced steroids are totally inactive. In the fertilized egg *in vivo*, the 5β-reduced steroids could be readily formed from cholesterol and other steroid precursors. (*b*) The 5β-reduced steroids are selectively bound to a protein in the cytoplasm but are not transported into the nucleus. (*c*) The synthesis of fetal haemoglobin is not controlled by the provision of fetal globin messenger RNA (mRNA); this is present all the time, but is not necessarily translated. (*d*) The 5β-reduced steroids activate protein factors required for the translation of fetal globin mRNA and this appears to be the rate-limiting step of the overall process. In the blastoderm, we therefore have all the ingredients for translational control by testosterone metabolites; this is in sharp contrast to the transcriptional model depicted in Fig. 4-3. The difference in basic mechanisms is emphasized in Fig. 4-8. These events in the blastoderm are not connected with sexual differentiation, and occur in early chick embryos irrespective of their sex. The 5β-reduced steroids have no androgenic activity in the strictest sense and all fertilized eggs contain identical amounts of stored cholesterol and steroid-metabolizing enzymes. In addition, sexual differentiation in the chick begins after 5 or 6 days of embryonic development; haemoglobin synthesis in the blastoderm occurs at least 3 days beforehand.

(*a*) Transcriptional mechanism: needs activated receptor, leading to mRNA synthesis

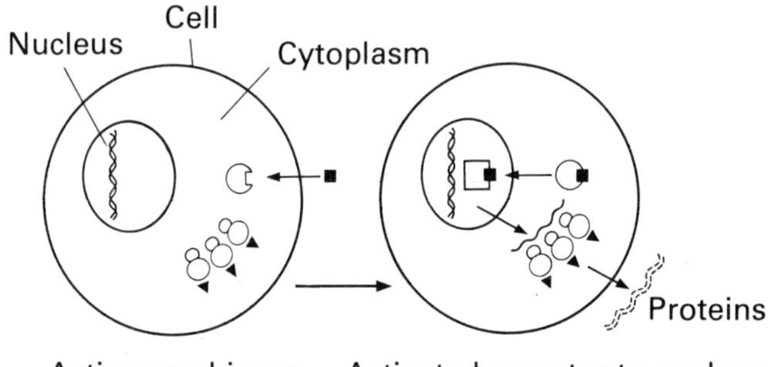

Active machinery, no mRNA

Activated receptor to nucleus, mRNA synthesized

(*b*) Translational mechanism: does not require activated receptor, rather direct stimulation of the machinery for protein synthesis

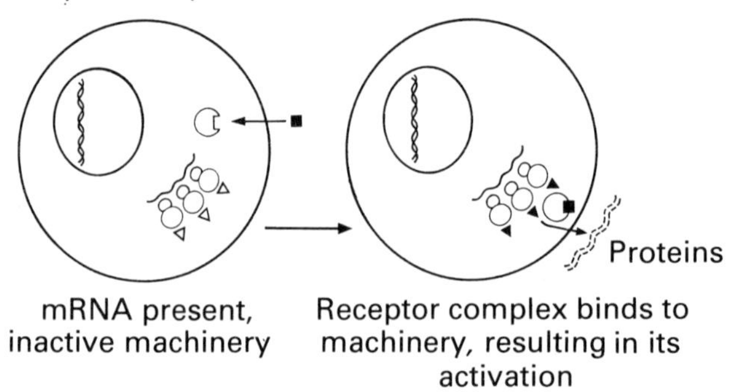

mRNA present, inactive machinery

Receptor complex binds to machinery, resulting in its activation

■ Hormone
Ⓒ Receptor
◻■ Activated receptor
〰 mRNA
△ Inactive factors for protein synthesis
▲ Activated factors
8 Ribosomes

Fig. 4-8. How a hormone can stimulate protein synthesis by different mechanisms: (*a*) the transcriptional (nuclear) mechanism and (*b*) the translational (cytoplasmic) mechanism.

Responses in the newborn

(*a*) *Liver enzymes.* In the rat and man, and maybe other species, enzymes involved in steroid metabolism in the liver are not the same in the two sexes. This sex-specificity may be related to the need to enhance the metabolism of the major steroid present in either sex, thus protecting the adult against excessive hormonal stimulation. A permanent change in the pattern of liver enzymes is brought about by neonatal secretion of testosterone, and this 'imprinting' can be prevented by anti-androgens; in the absence of androgens, the liver enzymes are more suited to the oestrogenic environment of the adult female. The molecular basis of imprinting is not clear, but it certainly involves pituitary hormones and the nuclear binding of testosterone and its metabolite, androstenedione. The involvement of androstenedione broadens its biological functions; it is already recognized as an important prehormone of testosterone (Book 3, Chapter 1) and also plays a part in controlling animal behaviour (Book 4, Chapter 2).

(*b*) *Accessory sex organs.* Injection of testosterone during the neonatal period has a profound effect on the subsequent growth of the seminal vesicles and coagulating glands during sexual maturation in adult male rats. In the adult rat, these two organs grow very much larger in size as a result of the earlier injection of testosterone, and they are said to be neonatally 'programmed', whereas the prostate and testis are not. The underlying mechanism is somewhat baffling because programming is insensitive to anti-androgens and presumably does not depend on a receptor system. The seminal vesicle and prostate originate from different anlagen and embryonic differentiation may leave them with a subtly different responsiveness to testosterone. We should appreciate that programming is an artificial response, involving testosterone in excess of that supplied by the neonatal testis. Also, whether programming can occur in many species or is peculiar to the rat is not clear. In sheep, androgenization of the male fetus produces an adult with very low levels of testosterone and LH, apparently because the negative feedback loop is reset.

(*c*) *Sexual differentiation of the brain.* We have known for some time that the sexual behaviour of rats, mice and hamsters can be irreversibly manipulated by changes in the hormonal environment immediately after birth, and similar effects can be produced by treatment before birth in guinea pigs, sheep, rhesus monkeys and human beings. Experimentally, sexual differentiation has been most widely studied in the rat. After neonatal castration, the male rat shows female behaviour in response to injections of testosterone. Conversely, when the newborn female rat is androgenized by a low dose of testosterone, she will show male behaviour in adulthood, a phenomenon known as neonatal androgenization. Sexual differentiation of the brain of the rat also regulates the pattern of secretion of gonadotrophins. In the female this is cyclical, since rising oestrogen levels cause an ovulatory discharge of LH, but in the male, this 'positive feedback' is abolished and gonadotrophin secretion is non-cyclical. The work of Frederick Naftolin in Montreal has recently suggested a possible explanation for neonatal androgenization, based on the following observations: (*a*) The hypothalamus of the newborn rat has an active aromatase enzyme system for converting androgens to oestrogens. (*b*) 5α-Dihydrotestosterone which cannot be converted into oestrogens by aromatase, cannot bring about neonatal androgenization. (*c*) There are both androgen and oestrogen receptors in the hypothalamus. (*d*) Neonatal androgenization can be induced by oestradiol-17β, and can be prevented by the anti-oestrogen, MER-25. These observations suggest that oestrogens may be the biologically active metabolites of androgens in the brain. Thus the neonatal testis secretes a short burst of testosterone which is not bound by α-fetoprotein and is converted into oestradiol-17β within the hypothalamus. This is then bound to the oestrogen receptors. Saturation of the oestrogen receptors promotes sexual differentiation of the brain and the process is predictably sensitive to anti-oestrogens.

TABLE 4-3. The biochemical responses of the ventral prostate of the castrated rat to daily injections of 2.5 mg of testosterone

Type of event and stimulation time (hours)	Biochemical responses
Initial (1–12)	Metabolism of testosterone Saturation of receptors Increase in RNA polymerase A Synthesis of ribosomal RNA and some messenger RNA Phosphorylation of nuclear proteins
Early (12–48)	Increase in RNA polymerase B Synthesis of mRNA Synthesis of polyamines Synthesis of membranes Assembly of polyribosomes Most proteins, including enzymes for secretion, now synthesized
Late (48–96)	Increase in all enzymes associated with DNA replication, especially DNA polymerase Synthesis of histones Cell division Growth

Responses at puberty

(*a*) *Secondary sexual characteristics and external genitalia.* The initial trigger for the onset of puberty is a decreased sensitivity of the hypothalamus to steroid feedback, resulting in an increased secretion of gonadotrophins, an increased synthesis of testosterone, and thus growth of the external genitalia and development of secondary sexual characteristics. When this growth is completed, the male animal is sexually mature. The model depicted in Fig. 4-3 is applicable in principle to all accessory sex glands and is based essentially on the selective formation and binding of 5α-dihydrotestosterone. Examples of the biochemical processes stimulated by androgens are given in Table 4-3. There are three categories, listed as initial, early and late events. Despite differences in their temporal appearance after androgenic stimulation, all of these processes are closely integrated. All male tissues grow rapidly in this period, but the final end products are different. In the testis, spermatozoa are produced and the accessory sex glands synthesize copious secretions which are stored within the lumen of each distinct gland.

The biochemistry of the secretions of the rat prostate and seminal vesicle is currently the centre of much interest. In both accessory sex organs, it is now clear that up to 40 per cent of the total protein synthesized is for secretion into the male urogenital tract. Our studies indicate that these secretions contain two or three predominant proteins and these are of different structure and function in the seminal vesicle and prostate. These organ-specific proteins are coded for by specific poly(A)-rich mRNA molecules which represent 35 per cent of the total mRNA in these organs after androgenic stimulation. The synthesis of these secretory proteins in cell-free conditions is now routine and the precise mechanism by which androgens regulate their synthesis *in vivo* is under active investigation. With respect to function, the seminal vesicle secretion is responsible for forming the vaginal plug in the female rat after coitus; thus secretory proteins are sometimes called clotting proteins. The function of the secretory proteins of the prostate is less evident as the gland is not essential for fertility. Walter Heyns in Leuven has evidence that the secretion of the prostate is rich in steroid-building proteins, which he terms collectively PBP, somewhat akin to the androgen-binding protein (ABP) synthesized in the testis and to be described later in this chapter. Taken overall, this new information adds weight to the argument that androgenic responses in the accessory sexual glands are based on a major enhancement of genetic transcription, particularly in terms of mRNA synthesis *de novo*.

Certain limitations in the model presented in Fig. 4-3 should now be highlighted. From the work of Jean Wilson we know that metabolites of testosterone may be present in some organs only during the limited period when sexual maturity is established, because many enzymes show acute developmental changes. The 5α-reductase in rat testis and in the prostate of the bull and rabbit is particularly active during the growth of these organs, but then it disappears completely. Once these organs have reached the maximum size, their subsequent maintenance must depend solely on testosterone or metabolites other than 5α-dihydrotestosterone. One may correlate prostate size with 5α-reductase activity in many species. In relation to body weight, the rat's prostate is large and has a high 5α-reductase activity throughout life; by contrast, the bull's prostate is tiny and possesses 5α-reductase activity only transitorily. Developmental changes in androgen-metabolizing enzymes therefore have a profound influence on the nature of androgenic responses, and 5α-dihydrotestosterone is the main stimulus for growth and cell division.

The overall model of androgen action (Fig. 4-3) does not explain satisfactorily the specificity of androgenic responses. For example, the brain is responsive to androgens throughout life but never in terms of growth. Similarly the accessory sex glands each secrete unique proteins yet the glands are closely related to one another anatomically, and have an identical androgenic environment and seemingly identical receptor systems. Perhaps the receptor proteins really are tissue specific but there is no experimental evidence to support this idea at present. The other alternative is that the androgen receptor systems do not have ultimate control over the target cells and that other metabolic 'governors' exist. It could be argued that during differentiation in the embryo, androgen target cells are structurally modified in such a way that certain genes are completely repressed and thereafter they are permanently resistant to activation by the androgen receptor complex. This could be true in the brain for the genes normally controlling growth and cell division. In the accessory sex organs, the genes must be suppressed in an even more subtle manner as the androgenic responses are so tissue specific. In rat prostate and seminal vesicle, we do know that mRNA coding for secretory proteins is synthesized in a restricted and tissue-specific manner. Thus the concept that mRNA synthesis is the critical event in androgenic responses is valid, but clearly there are regulatory elements in androgen target cells other than just the receptor systems. Current evidence suggests that non-histone proteins regulate the genetic expression of DNA in chromatin and the individual character of these proteins is probably established during differentiation.

We should also consider the availability of genes for transcription by RNA polymerase during the process of hormonal stimulation. Expressed another way, are the genes responsible for initial responses (e.g. ribosomal RNA synthesis) available at all times for transcription, whereas those for the later responses (e.g. histone and DNA synthesis) are not? Current evidence supports this idea. Even in cells deprived of androgens, the ribosomal RNA genes are freely available but the histone genes are inaccessible to RNA polymerase until the cells are stimulated to divide by the administration of testosterone. The classical response to androgenic stimulation is growth; maybe cell division and the separation of the chromosomes are necessary for the expression of late, but not initial, responses in rat prostate.

(*b*) *Muscle.* Muscle is such a major proportion of the body weight of male animals that the anabolic effect of testosterone is important and will be discussed briefly. Research on other androgen target cells has led to little further understanding of these anabolic effects than we had 30 years ago when they were first described. Certainly the anabolic response does not involve the metabolism of testosterone, as muscle does not possess the appropriate enzyme systems. In addition, cultures of isolated muscle cells grow faster in the presence of testosterone, but potential metabolites, such as 5α-dihydrotestosterone, have no effect. Even the presence of a receptor mechanism in muscle remains the centre of considerable debate, and even if it does exist, the binding machinery is much less efficient than that in the prostate. Muscle has an extremely high cytoplasmic : nuclear ratio and there is no evidence at all for the nuclear binding of testosterone. These facts, together with the inexorable slowness of the anabolic response, are not consistent with our working model and the interaction between testosterone and muscle remains an enigma. The situation would be helped by studies on the binding of radioactive anabolic agents, such as stanazol or norbolethone, but surprisingly these experiments have not been done.

A well-guarded secret, occasionally exposed by random urine tests, is that many international athletes take anabolic steroids in order to build up their muscles for field events needing brute strength. But studies on weight lifters and shot putters indicate that this practice may be dangerous. Anabolic steroids work slowly; they must be taken at high doses for protracted periods. Infertility, muscular degeneration and kidney disorders may accompany the long-term use of anabolic steroids; any of the consequences are a heavy personal price to pay for a winner's medal.

(*c*) *Mouse salivary gland.* The submaxillary gland in the mouse is sexually dimorphic. In the normal male mouse and in testosterone-treated females, the gland assumes a 'male-type' morphology and biochemistry, resulting in the secretion of male-specific proteases and other proteins, including the nerve growth factors, so-called because of their interesting ability to stimulate the growth of isolated neurones and other nerve cells in cell culture. The growth of the submaxillary gland is an interesting example of a pronounced morphological effect of testosterone in a 'non-sexual' gland and is illustrated in Fig. 4-9.

(*d*) *Mouse kidney.* In mice of similar age, the kidneys in the male are larger and histologically distinct from those of the female; this difference is attributable to circulating androgens. Elegant studies by Kenneth Paigen in Buffalo on the induction of β-glucuronidase by androgens illustrate the real complexity of the control mechanisms in higher animals. The genes that

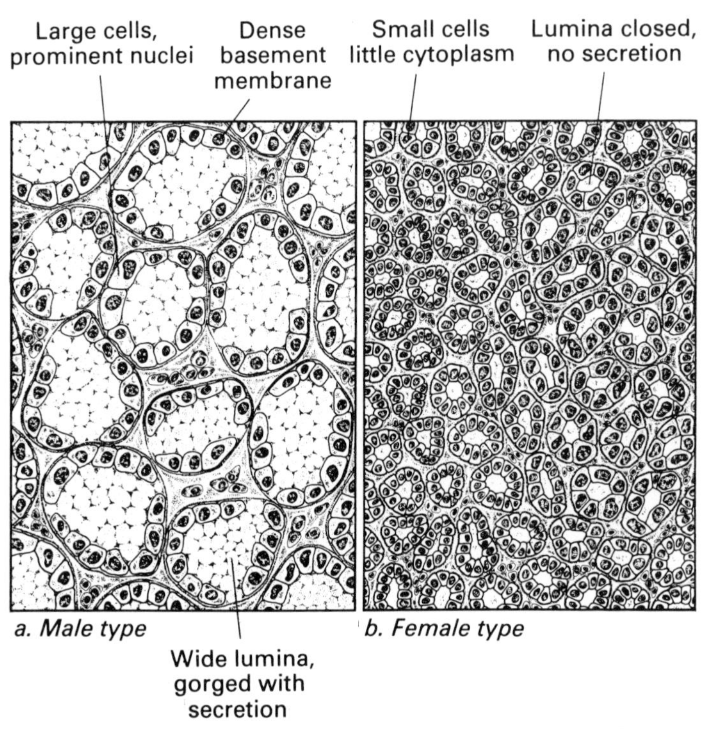

Fig. 4-9. Sexual dimorphism in mouse submaxillary salivary gland. The male (secretory) form (*a*) is present in testosterone-treated females and normal males, and the female (non-secretory) type (*b*) is present in females and castrated males.

regulate the synthesis of this hydrolase enzyme in mouse kidney have now been identified, but other genes are also involved. These regulate other aspects of the response, including the molecular size of the enzyme, its intracellular location, its inducibility by androgens and its storage. At least six genes are needed in all and Paigen has placed these in one of four categories, depending on their function. Let us examine this interesting system in a little more detail.

The basic subunit of β-glucuronidase is a polypeptide of molecular weight (M_r) 70000, but the six functional forms of the enzyme are tetramers, M_r range 260000–470000, and may be

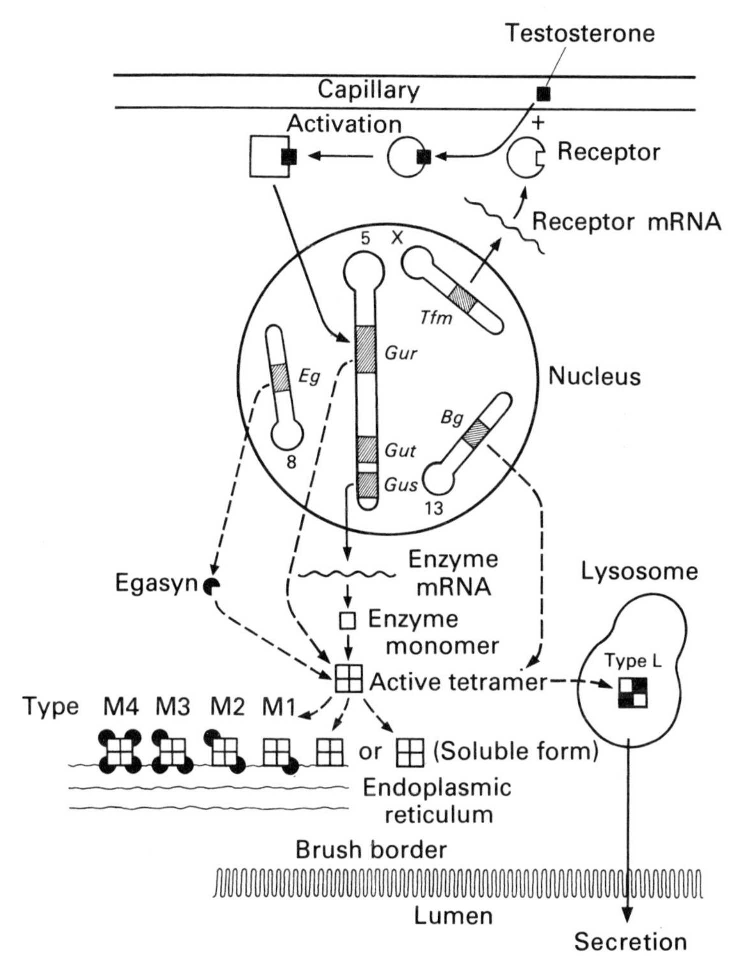

Fig. 4-10. Genetic control of β-glucuronidase in mouse kidney. The important genetic loci on the chromosomes are indicated with hatching. Full details are given in the text. Adapted from Paigen *et al.* *J. Cell Physiol.* **85**, 379 (1975).)

found either in the lysosomes or the cell soluble fraction, or else bound to membranes of the endoplasmic reticulum. These may be summarized as follows: type L (lysosomes, M_r 260000), type M (four species: microsomes, M_r 310000–470000 and type × (soluble fraction and microsomes, inducible by androgens, M_r 260000, but distinct from L). The four species of M enzyme contain varying amounts of another protein, egasyn, which enables the enzyme tetramer to bind to the membranes of the endoplasmic reticulum. The six genetic loci involved, together with their classification and function, are as follows. (i) *Gus*: structural gene controlling the synthesis of the mRNA for the enzyme monomer. (ii) *Tfm*: structural gene controlling the mRNA for the androgen receptor protein; the X form of enzyme is not induced by androgens in *Tfm* mutants. (iii) *Gur*: regulatory gene controlling the androgenic induction of the enzyme; the androgen receptor complex is believed to bind to and activate this gene. (iv) *Gut*: temporal gene controlling the appearance of the enzyme during development. (v) *Eg*: processing gene regulating the synthesis of egasyn, the 'binder' for the enzyme. (vi) *Bg*: processing gene controlling the entry of enzyme into the lysosomes and its secretion from the kidney. These concepts are summarized in Fig. 4-10. Apart from its obvious scientific merit, this work further underlines the usefulness of mutants and inbred strains of mice in biological research.

(*e*) *Testis*. As one who has spent much of his research life working on the prostate and other accessory sexual glands, it shames me to admit that I may have misjudged my career; there is no doubt that the principal biological role of testosterone is to maintain spermatogenesis. The testis itself is a complex organ, subject to hormonal control by LH, FSH and testosterone, and spermatogenesis is a remarkably sophisticated process on which I am no authority. Studies by Vidar Hansson in Oslo, Anthony Means in Houston and Frank French in Chapel Hill have collectively given new insights into the hormonal control of testicular function. The two gonadotrophins are now known for certain to work on different cell types in the testis, but both share a common mechanism in binding at specific membrane receptors in their target cells, thereby promoting the synthesis of cyclic AMP. LH interacts specifically with the Leydig cells, enhancing the conversion of cholesterol into testosterone. FSH interacts

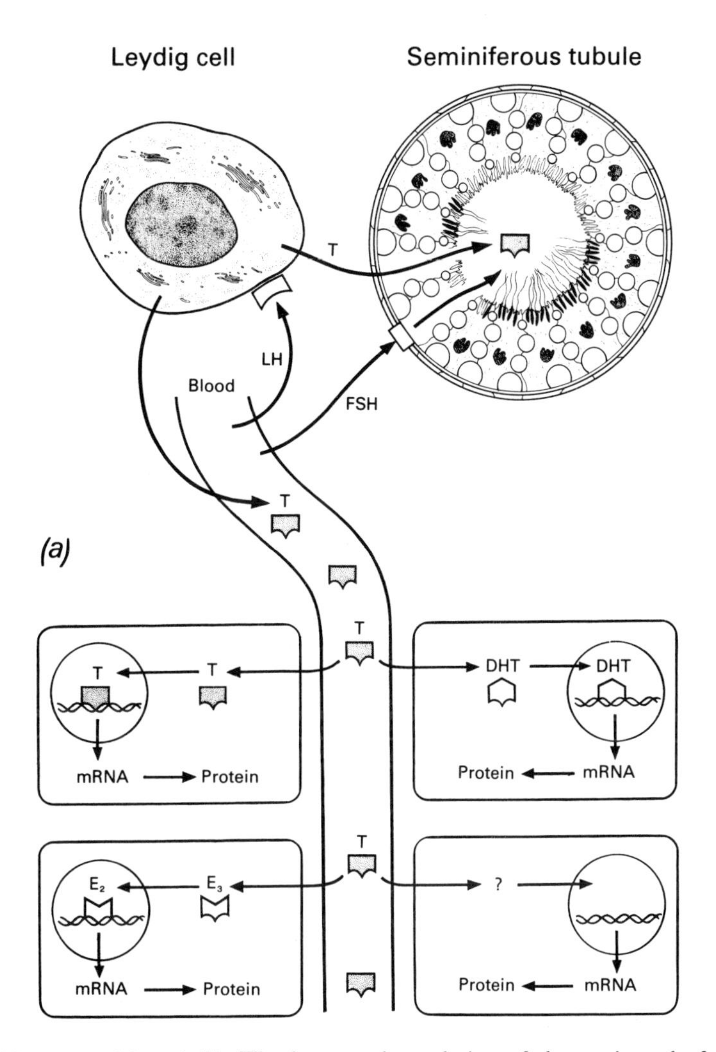

Fig. 4-11. (*a*) and (*b*). The hormonal regulation of the testis and of functions in other tissues. ABP, androgen-binding protein; AC, adenylate cyclase; C, cholesterol; DHT, 5α-dihydrotestosterone; E_2, oestradiol; MC, myoid cell; R, receptor.

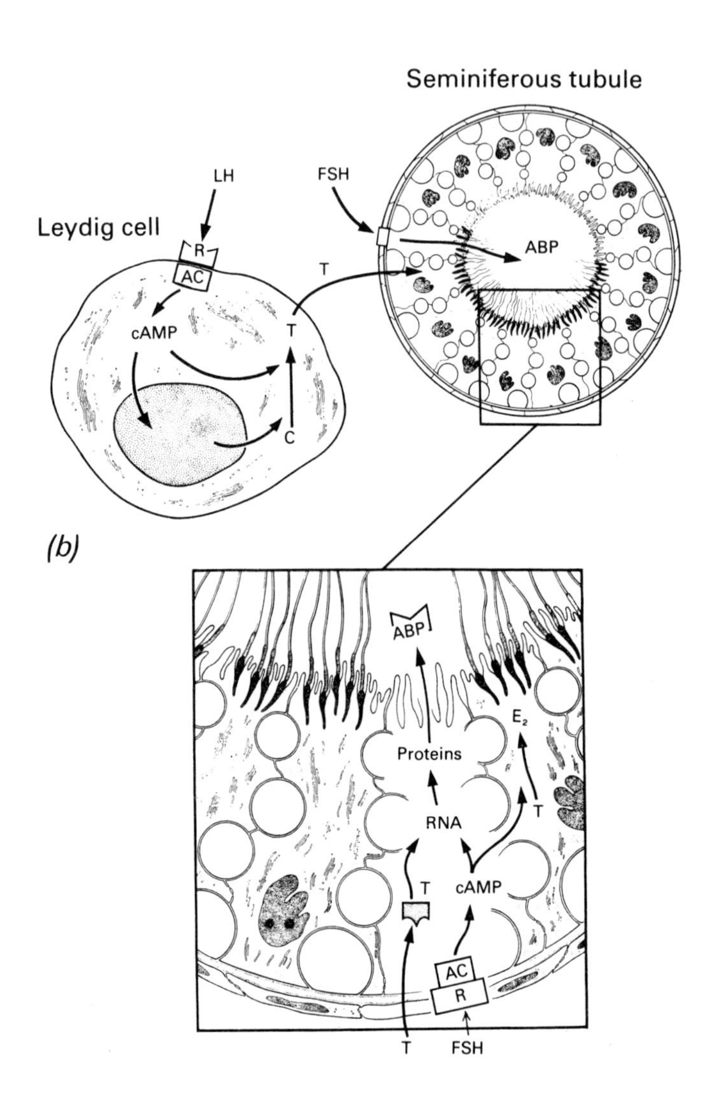
Seminiferous tubule
LH
FSH
Leydig cell
R
AC
ABP
T
cAMP
T
C
(b)
ABP
E2
Proteins
T
RNA
T
cAMP
AC
R
T
FSH

specifically with the Sertoli cells, and together with testosterone, regulates spermatogenesis and the synthesis of certain proteins. Testosterone is bound by an androgen receptor system seemingly identical to that described in other androgen target cells; the primary location of these androgen receptors remains a little uncertain, but is probably in the Sertoli cells of the seminiferous tubules. How FSH and testosterone regulate spermatogenesis still remains to be established unequivocally, but we do know that, in combination, they promote the synthesis of the androgen-binding protein (ABP) in Sertoli cells, both *in vivo* and in culture *in vitro*. ABP synthesis is now widely used as a marker of Sertoli cell function and LH plays no part in this process. One would particularly like to know how the nuclear binding of testosterone in the testis influences the target organ response in ABP synthesis. Work involving ligation of the efferent ducts has shown that ABP is secreted into the fluid of the seminiferous tubules and may be recovered from the epididymis; indeed, it was first discovered there. On current evidence, ABP provides the seminiferous fluid with the means to accumulate androgens from their site of synthesis in the Leydig cells; the maintenance of high local concentrations of testosterone in the seminal fluid and epididymis appears necessary for the viability and maturation of the spermatozoa. The hormonal regulation of the testis is illustrated in Fig. 4-11.

I have attempted to summarize the mechanism of action of androgens. It is a wide topic and although light has been shed on some areas, others still remain in the dark. Much has been said about the most popular model systems for studying the actions of androgens, the rat ventral prostate, but this has left the more exotic responses unexplored. The growth and shedding of antlers in male deer, the patterns of social behaviour, the thickening of the human vocal cords and even the leg-cocking action of the male dog – these remain among the more mysterious effects of testosterone. The mechanisms underlying such events will surely be elucidated some day. 'Courage, mon brave!' is the rallying cry – and should not this be shouted in a really deep voice?

SUGGESTED FURTHER READING

The mechanism of action of androgens. W. I. P. Mainwaring. *Monographs on Endocrinology*, vol. 10. New York; Springer-Verlag (1977).

The effects of androgens on the complexity of messenger RNA from rat prostate. M. G. Parker and W. I. P. Mainwaring. *Cell* **12**, 401 (1977).

Testosterone regulates the synthesis of major proteins in rat ventral prostate. M. G. Parker, G. T. Scrace and W. I. P. Mainwaring. *Biochemical Journal* **170**, 115 (1978).

Androgen-dependent synthesis of basic secretory proteins by the rat seminal vesicle. S. J. Higgins, J. M. Burchell and W. I. P. Mainwaring. *Biochemical Journal* **158**, 271 (1976).

The regulation of haemoglobin synthesis in cultured chick blastoderms by steroid related to 5β-androstane. R. A. Irving, W. I. P. Mainwaring and P. M. Spooner. *Biochemical Journal* **154**, 83 (1976).

Metabolism of testosterone and actions of metabolites on prostate glands in organ culture. E.-E. Baulieu, I. Lasnitski and P. Robel. *Nature* **219**, 1115 (1968).

The intranuclear metabolism of testosterone in the accessory organs of reproduction. J. D. Wilson and R. E. Gloyna. *Recent Progress in Hormone Research* **26**, 309 (1970).

Male pseudohermaphroditism: the complexities of male phenotypic development. J. Imperato-McGinley and R. E. Peterson. *American Journal of Medicine* **61**, 251 (1976).

X-linked gene for testicular feminization in the mouse. M. F. Lyon and S. G. Hawkes. *Nature* **227**, 1127 (1970).

Normal spermatozoa from androgen-resistant germ cells of chimaeric mice and the role of androgen in spermatogenesis. M. F. Lyon, P. H. Glenister and M. L. Lamoreux. *Nature* **258**, 620 (1975).

The formation of estrogens by central neuroendocrine tissues. F. Naftolin, K. J. Ryan, I. J. Davies, V. V. Reddy, F. Flores, A. Petro, M. Kuhn, R. J. White, Y. Takaoka and L. Wolin. *Recent Progress in Hormone Research* **31**, 295 (1975).

The molecular genetics of mammalian glucuronidase. K. Paigen, R. T. Swank, S. Tomino and R. E. Gamshow. *Journal of Cell Physiology* **85**, 379 (1975).

Alterations in testicular morphology and function in rabbits following active immunization with testosterone. E. Nieschlag, K.-H. Usadel, O. Schwedes, H. Kley, K. Schoffling and H. L. Kruskempar. *Endocrinology* **92**, 1142 (1973).

5 The oestrogens

E. V. Jensen

E. V. Jensen

DESCRIPTION, OCCURRENCE AND ACTIONS OF OESTROGENS

The oestrogens comprise a class of steroid hormones of principal but not exclusive importance to reproductive processes in the female. They are characterized chemically by an aromatic A-ring bearing a phenolic group in position 3 of the steroid nucleus (Fig. 5-1). During the reproductive years, the principal oestrogen in

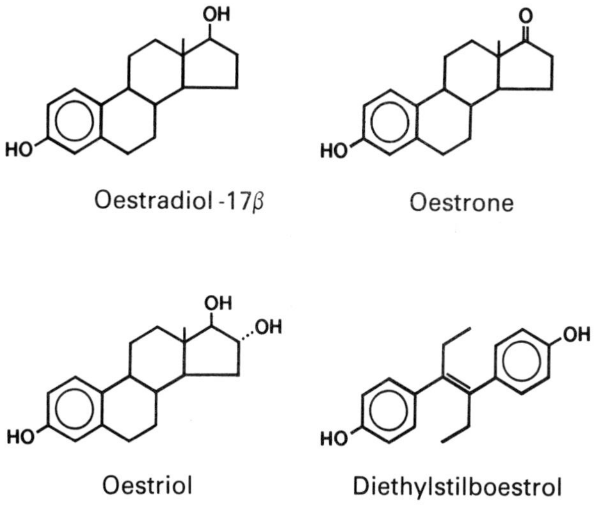

Oestradiol-17β Oestrone

Oestriol   Diethylstilboestrol

Fig. 5-1. Structures of some natural and synthetic oestrogens

women is oestradiol-17β, produced from cholesterol in the ovary under stimulation by gonadotrophic hormones from the anterior pituitary (Chapter 2). In the liver and elsewhere, circulating oestradiol is converted in large part to the less active substance, oestrone, which, along with oestriol derived from it, represents a major urinary excretion product. In postmenopausal women, where ovarian production of steroids has abated, the principal oestrogen is oestrone, derived from the peripheral conversion of 4-androstene-3,17-dione originating in the adrenal glands. During human pregnancy, the feto-placental unit produces substantial amounts of both oestradiol and oestrone, as well as oestriol, especially during the last trimester. In men, oestradiol, in amounts roughly one-fifth that produced in non-pregnant women, is secreted by the Leydig cells of the testis.

The physiological actions of the natural steroid oestrogens all can be reproduced by various types of non-steroidal phenolic substances. The most familiar of these are synthetic derivatives of stilbenes, such as diethylstilboestrol (Fig. 5-1) and its saturated analogue, hexoestrol. Certain triarylethylenes (tri-*p*-anisylchloroethylene) and polycyclic lactones also are active oestrogens, although of somewhat lower potency. Oestrogenic lactones (coumestrol, zearalenone) are found in certain plants, often in quantities sufficient to cause disruption of endocrine balance and loss of fertility in farm animals ingesting them.

The oestrogenic hormones exert effects in a large number of different tissues. Their principal physiological action is usually considered to be stimulation of growth and development of the female reproductive organs, in particular the uterus, vagina and mammary glands. The ovarian production of oestrogens varies periodically (menstrual cycle in primates, oestrous cycle in lower mammals) so that in addition to maintaining the reproductive tissues in their adult form, oestrogens cause intermittent episodes of intensive stimulation, especially in the uterine endometrium and vaginal epithelium (Book 3, Chapter 3). The oestrogens also promote or influence a number of other physiological processes, including sexual behaviour, development of female secondary sex characteristics (pelvic enlargement, generalized fat deposition), formation of bone matrix by osteoblasts, and epiphyseal closure in growing bones. Nidation of the fertilized ovum and proper development of the embryo depend on the combined action of oestrogen and progesterone (Book 3, Chapter 4). As in the case of other gonadal hormones, oestrogens exert an inhibiting action on the hypothalamus and anterior pituitary gland, thereby effecting feed-back regulation of the level of gonadotrophins that stimulate their production by the ovary (Chapters 1 and 2). Oestrogens also induce the synthesis of progesterone receptors in female reproductive tissues, which explains the long recognized need for oestrogen prestimulation or 'priming' in order for these tissues to respond to progesterone. In certain non-mammalian species an important action of oestrogens is to promote the synthesis of proteins involved in egg production, such as ovalbumin in the chick oviduct and vitellogenin in the frog liver.

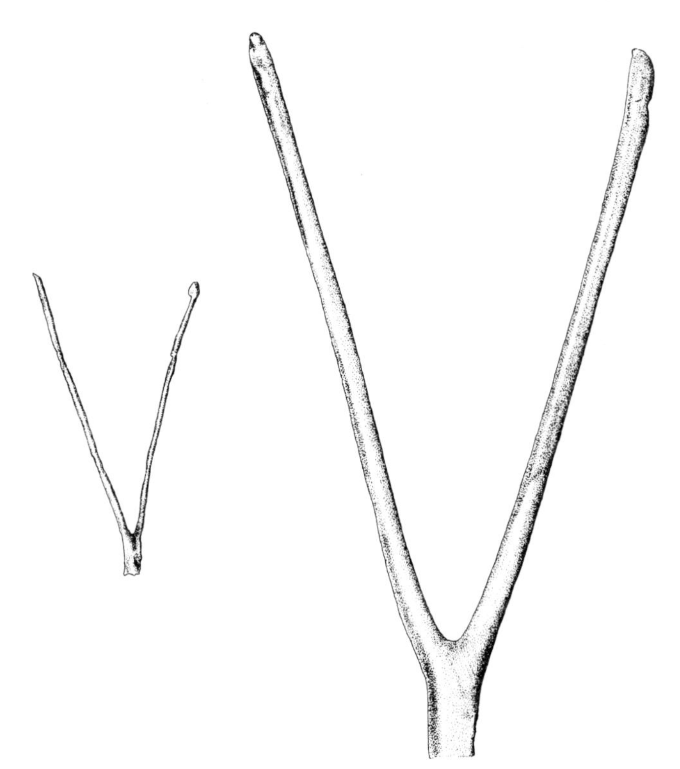

Fig. 5-2. Uterus of immature rat before and 1 day after a series of four daily subcutaneous injections of 100 ng oestradiol in saline

The biological responses that have most often been employed as indicators of oestrogenic activity are the induction of uterine growth in immature or ovariectomized mice or rats (Fig. 5-2), and the stimulation of metaplasia of the vaginal epithelium (Fig. 5-3) with the formation of keratin and release of characteristic 'cornified' cells (Allen-Doisy test). For investigations of the biochemical mechanism of oestrogen action, the immature rat uterus has proved especially useful; the existence of steroid hormone receptors and the general pattern of hormone–receptor interaction in target cells were originally elucidated with this experimental system.

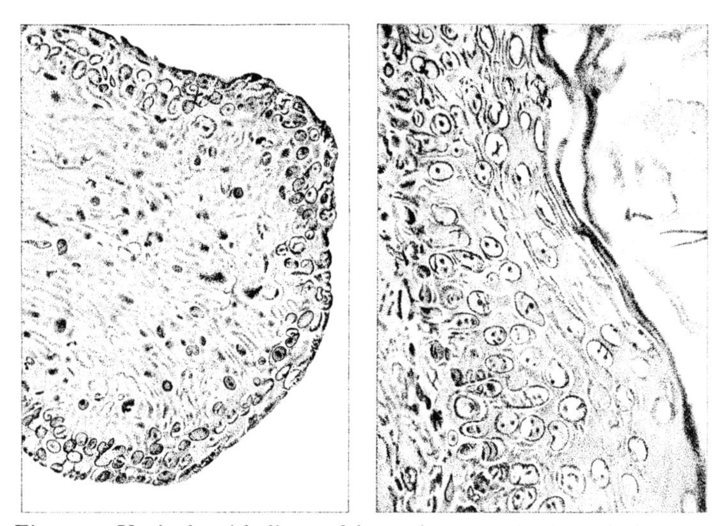

Fig. 5-3. Vaginal epithelium of hypophysectomized rat before and 1 day after a series of 6 daily injections of 100 ng oestradiol in sesame oil

MECHANISTIC APPROACHES TO OESTROGEN ACTION

In contrast to knowledge as to what the oestrogenic hormones are and what they do, our understanding of how they do it is far from complete. Still, the past two decades have seen remarkable progress towards the elucidation of biochemical mechanisms of steroid hormone action. This has come from two general lines of investigation, both beginning in the late 1950s. One approach, originating with the studies of Gerald Mueller and his associates in Wisconsin, involves detecting early effects of administered oestrogen on biochemical reactions in target cells of hormone-derprived animals – usually by determining the rates of incorporation of labelled precursors into tissue constituents – and the effects of specific inhibitors on these biosynthetic processes. The other approach, initiated by the synthesis of [^{3}H]hexoestrol by Glascock and of [^{3}H]oestradiol in our own laboratory, consists of determining the biochemical fate of the radioactive hormone as it exerts its action in target cells. In one case the question has been 'what does the hormone do to the tissue?'; in the other, 'what does the tissue do with the hormone?'

The first approach, with contributions from the laboratories of Hamilton, Tata, Gorski, Segal, Smellie, O'Malley, Schimke, Spelsberg, Glasser, Warren and others, has established that a very early effect of administered oestrogen is to increase the rate of synthesis and/or processing of various types of RNA. From the second approach has come the recognition that the oestrogen-dependent tissues contain characteristic hormone-binding proteins or receptors, that most of the oestrogen taken up by target cells becomes localized in the nucleus, and that it is the oestrogen–receptor complex, rather than the hormone itself, that elicits the biochemical response. Studies of oestradiol binding to macromolecular receptors, with important contributions from many investigators including Stone, Gorski, King, Terenius, Eisenfeld, Talwar, Jacobson, Jungblut, Baulieu, Erdos, Korenman, Stumpf, Notides, Clark, Puca, Wotliz, Sherman, Alberts, Siiteri, Rochefort, Wittliff, McGuire, Pasqualini, Katzenellenbogen, De Sombre and Ruh, have elucidated the general pattern of oestrogen–receptor interaction in target cells.

Correlation of information obtained from both types of approach has led to an overall picture of oestrogen action (Fig. 5-4). The hormone enters the target cell, apparently by passive diffusion, and binds to an extranuclear receptor protein (Rc)

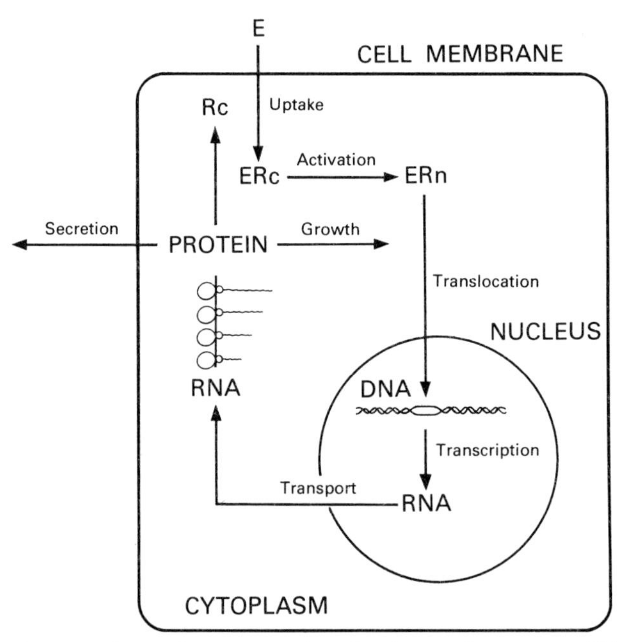

Fig. 5-4. General representation of pathway of oestrogen (E) interaction and biochemical responses in target cells. Rc, receptor protein (cytosol form); Rn active form (nuclear) of receptor protein.

inducing its conversion to an active form (Rn), which, unlike the native receptor, has a strong affinity for chromatin. The activated oestrogen–receptor complex (ERn) is translocated to the nucleus, where it binds in the chromatin and in some way enhances processes associated with the production and utilization of messenger and preribosomal RNAs needed for the synthesis of constituent, enzymic and secretory proteins, including the receptor itself. The following sections present an overview of the key experimental observations leading to the concept of the two-step translocation mechanism, originally elucidated for the oestrogens but applicable as a general model for the action of all classes of steroid hormones in their respective target cells.

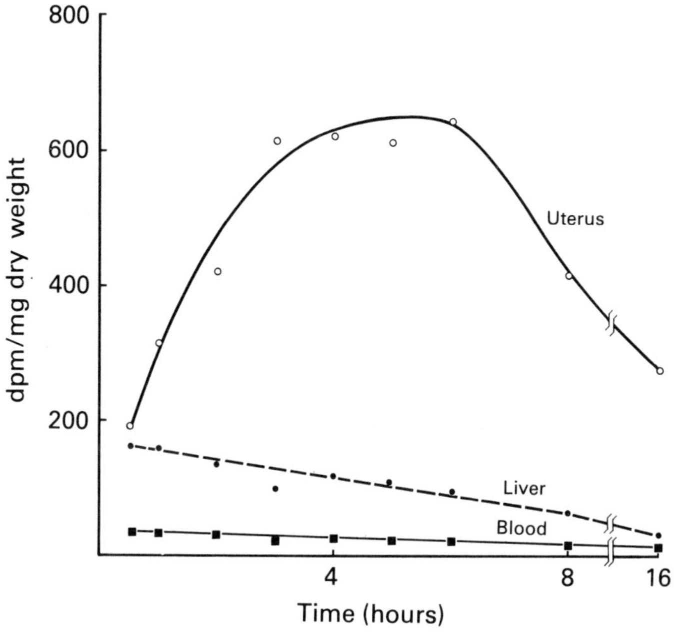

Fig. 5-5. Concentration of radioactivity in tissues of 23-day-old rat after single subcutaneous injection of 10 ng (1.21 μCi) [^{3}H]oestradiol in sesame oil. (From E. V. Jensen and H. I. Jacobson. *Recent Progr. Hormone Res.* **18**, 387 (1962).)

UPTAKE OF OESTROGEN HORMONES BY TARGET TISSUES

The fact that the female reproductive tissues contain a characteristic oestrogen-binding component, called oestrogen receptor or oestrophilin, was first indicated by their striking ability to take up and retain [^{3}H]oestrogens after the administration of physiological doses *in vivo* (Fig. 5-5) and, later, on exposure of excised tissues to the hormone *in vitro* (Fig. 5-6). It is now recognized that most, if not all, mammalian tissues contain small amounts of oestrogen receptor and that the unique characteristic of the hormone-dependent tissues is the magnitude of their oestrophilin content. Oestradiol was found to combine reversibly with the receptor and to initiate growth of the immature rat

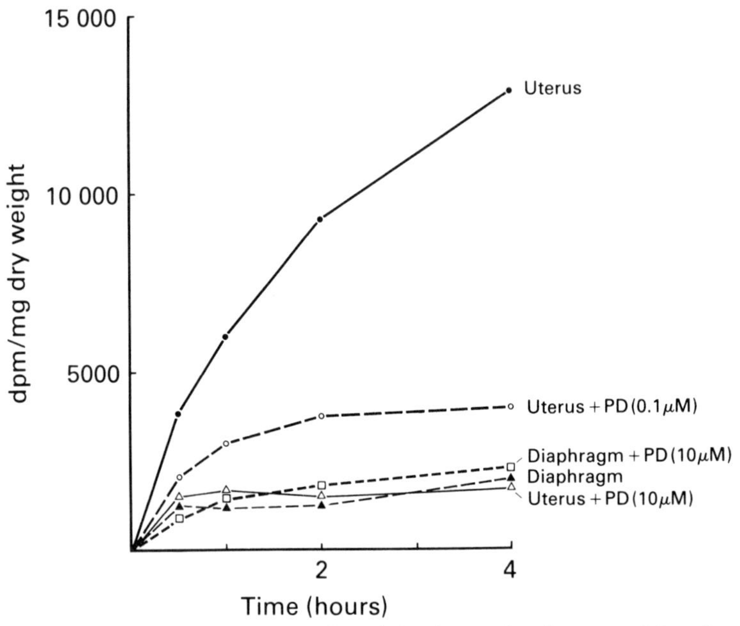

Fig. 5-6. Concentration of radioactivity in uterine horns and hemi-diaphragms of immature rats after exposure to 0.12 nM [^{3}H]oestradiol at 37 °C in Krebs–Ringer–Henseleit glucose buffer, pH 7.3, in the presence and absense of the oestrogen antagonist, Parke Davis CI-628 (PD). (From E. V. Jensen *et al. Gynecol. Invest.* **3**, 108 (1972).)

uterus without itself undergoing chemical change, suggesting that the action of the hormone involves its influence on macromolecules, rather than participation in reactions of steroid metabolism as had once been assumed.

The specific uptake and retention of oestradiol by target tissues, both *in vivo* and *in vitro*, is inhibited by a class of oestrogen antagonists that are themselves very weak oestrogens, but which prevent the striking uterotrophic action of the natural hormone. As illustrated in Fig. 5-6, these substances, which include clomiphene, nafoxidine, Parke Davis CI-628 and tamoxifen (Fig. 5-7), provide a useful means for distinguishing specific binding of hormone to receptor from the non-specific binding that oestradiol shows with all tissues or with various

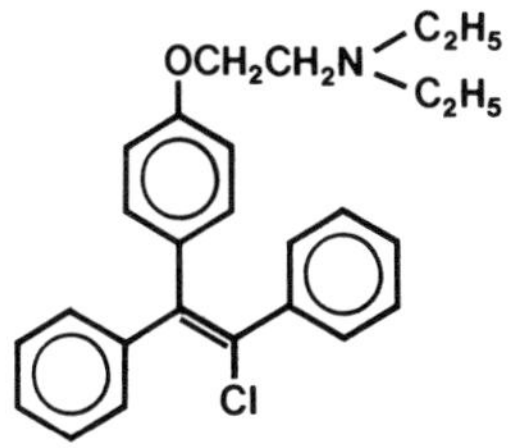

MRL-41 Clomiphene

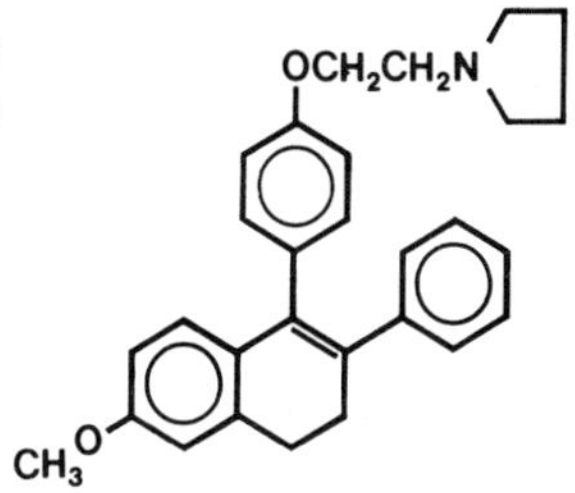

U-11,100 Nafoxidine

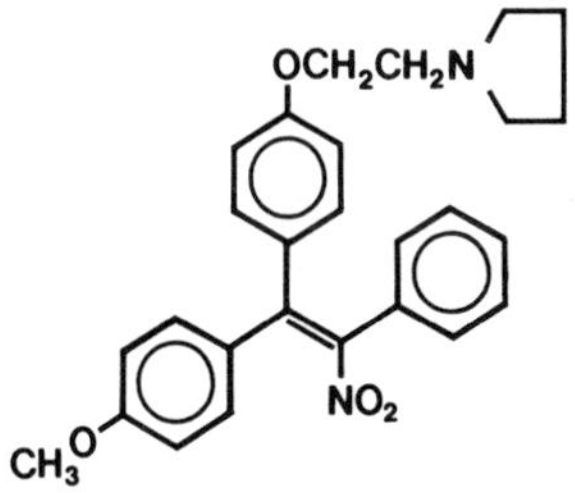

PD CI-628

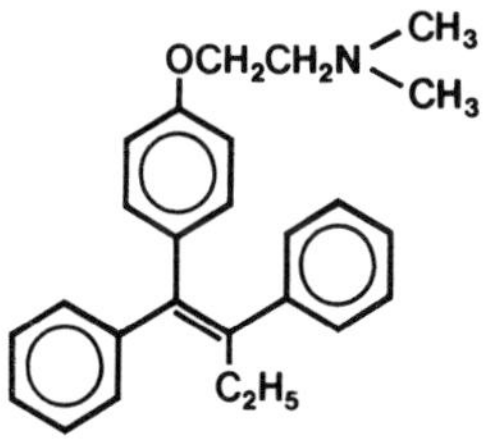

ICI-46 474 Tamoxifen

Fig. 5-7. Structures of oestrogen antagonists

macromolecules in broken cell systems. The correlation observed between the reduction in hormone incorporation and the inhibition of uterine growth when different amounts of nafoxidine are administered along with oestradiol to the immature rat, first provided evidence that binding of hormone to receptor actually is involved in its biological action. In contrast, actinomycin D and puromycin, substances that prevent the growth response to oestradiol, show no inhibition of the characteristic uptake and retention of hormone. This suggests that the binding of oestradiol to receptor is an early step in the uterotrophic process, initiating a sequence of biochemical events that can be blocked at later stages by these inhibitors of RNA and protein synthesis.

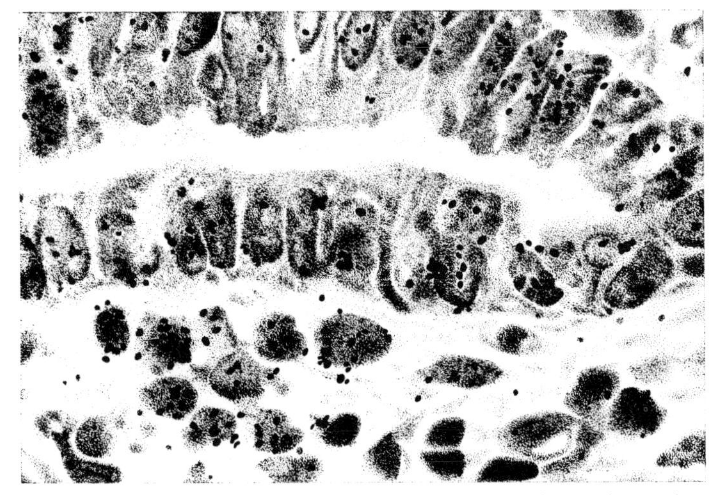

Fig. 5-8. Autoradiograph of frozen section of rat uterine endometrium excised 2 hours after subcutaneous injection of [^{3}H]oestradiol in saline. Courtesy of Dr W. E. Stumpf.

OESTROGEN–RECEPTOR INTERACTION

Hormone–receptor complexes

When uterine homogenates from oestradiol-treated rats are subjected to differential centrifugation, the incorporated steroid appears in two cellular fractions. Most of the hormone (70–80 per cent) is found in the nuclei, with a smaller amount present in the high-speed supernatant or cytosol fraction. The predominance of nuclear binding, controversial in earlier reports, was confirmed by autoradiographic studies using a dry-mount procedure that minimizes steroid translocation during tissue processing (Fig. 5-8). Similar nuclear localization of hormone is seen when excised uteri are exposed to oestradiol at physiological temperatures *in vitro*.

The oestradiol taken up by rat uterus is associated with a different form of the receptor in the cytosol than in the nucleus. The application by Toft and Gorski of ultracentrifugation in sucrose density gradients for characterizing oestrogen–receptor

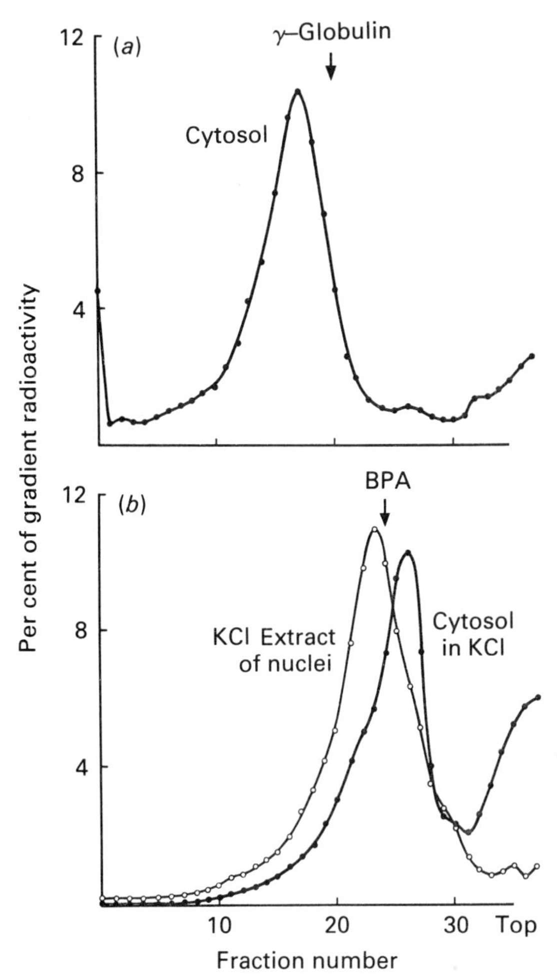

Fig. 5-9. Sedimentation patterns of radioactive oestradiol–receptor complexes of rat uterine cytosol and nuclear extract (400 mM KCl) from uteri of immature rats excised 1 hour after the subcutaneous injection of 100 ng (20.8 μCi) [^{3}H]oestradiol in saline. To saturate its receptor capacity the cytosol fraction was made 5 nM with additional [^{3}H]oestradiol. γ-Globulin and BPA indicate positions of bovine immunoglobulin (7.0 S) and bovine plasma albumin (4.6 S) markers. Gradients are: (*a*) 10–30 per cent sucrose without added salt; (*b*) 5–20 per cent sucrose containing 400 mM KCl. (From E. V. Jensen and E. R. DeSombre. *Science* **182**, 126 (1973).)

complexes provided a valuable means for distinguishing between different modifications of the receptor. By this technique the radioactive hormone in the cytosol is found to sediment as a discrete band with a coefficient close to 8 S (Fig. 5-9*a*). In salt concentrations greater than 0.2 M, the 8 S complex is reversibly dissociated into subunits that sediment at about 4 S, just behind bovine plasma albumin (Fig. 5-9*b*). The oestradiol bound in the nucleus can be solubilized, unaccompanied by DNA, by extraction with 0.3 M or 0.4 M KCl to yield an oestradiol–receptor complex which, in the presence of salt, sediments at about 5 S, slightly faster than bovine plasma albumin. As shown in Fig. 5-9*b*, the nuclear complex is readily distinguished from the cytosol complex by careful ultracentrifugation in salt-containing sucrose gradients; this difference in sedimentation rates provided the first criterion for recognizing the important phenomenon of receptor activation.

The 8 S oestradiol–receptor complex, or its 4 S subunit, forms directly in the cold when oestradiol is added to the cytosol fraction of uteri not previously exposed to hormone. Thus, the receptor content of the cytosol is easily estimated by adding sufficient [^{3}H]oestradiol to saturate the binding sites and determining the radioactivity present in the 8 S sedimentation peak. This interaction of oestradiol with cytosol receptor is prevented by the presence of anti-oestrogens, such as nafoxidine or CI-628. Though noncovalent, the binding of oestradiol to receptor proteins of uterine tissue is remarkably strong; association constants varying from 10^9–10^{12} M^{-1} have been reported for the cytosol complex. This tight binding appears to result from a very slow rate of dissociation; once formed, the complex does not readily lose oestradiol in the cold except by receptor decomposition.

From the sensitivity of their complexes to proteases but not to nucleases, the oestrogen-binding substances of both cytosol and nucleus appear to be mainly protein in composition. They are rather unstable in crude extracts, tending to aggregate and to decompose during storage or attempted purification. Addition of Ca^{2+} to the salt-dissociated complex of uterine cytosol yields a stabilized 4.5 S form of the binding unit that is resistant to aggregation and does not revert to the 8 S state on removal of salt. This stabilization appears to result from the activation by Ca^{2+} of an enzyme, present in uterine cytosol, that acts on the receptor protein to destroy its ability to aggregate in media of low ionic strength. The physiological significance of this phenomenon, if any, is not known. On removal of salt, the 5 S complex in nuclear extracts aggregates to a 8–9 S form, but loses this property on aging or partial purification.

Because of its instability and tendency to aggregate, as well as its surprising resistance to elution from affinity chromatography columns, oestrophilin has proved difficult to purify. Recently, purification to apparent homogeneity has been reported for both the cytosol and nuclear oestradiol–receptor complexes of calf uterus, and nuclear material approaching purity has been used to prepare antibodies to oestrophilin, as discussed in a later section.

Receptor translocation

A major advance in the understanding of the interaction of steroid hormones with target cells came with the recognition that the oestradiol–receptor complex of the uterine nucleus is derived from the cytosol complex by a temperature-dependent process in which association with the hormone causes the extranuclear receptor to localize in the nucleus (Fig. 5-10). It was later shown that the temperature-dependent aspect of this phenomenon is the hormone-induced conversion of the native oestrophilin to an active form that can bind to chromatin. Although this two-step receptor translocation mechanism is not proved with absolute certainty, it is supported by many different types of experimental evidence (Table 5-1).

A relation between the two intracellular sites of oestrogen localization was first indicated by observations that a given dose of nafoxidine *in vivo* inhibits cytosol and nuclear binding of

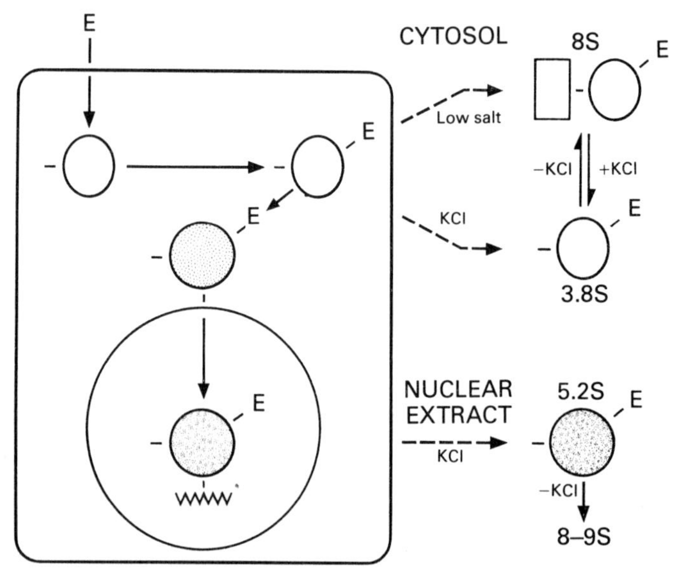

Fig. 5-10. Schematic representation of interaction pathway of oestradiol in target cell. Diagram at left shows oestradiol (E) combining with oestrophilin to induce receptor activation followed by translocation of the transformed complex to bind to chromatin in the nucleus. Diagram at right indicates sedimentation properties of complexes extracted from the cell after homogenization

TABLE 5-1. Evidence for two-step receptor translocation mechanism

1. Cytosol dependence of receptor complex formation in isolated nuclei.
2. Temperature-dependent shift of extranuclear to nuclear binding.
3. Oestrogen-induced depletion of extranuclear receptor *in vivo*.
4. Oestrogen-induced conversion of cytosol oestrophilin from native (4 S) to nuclear (5 S) form.
5. Binding and stimulation of RNA synthesis by activated extranuclear complex in isolated target cell nuclei.
6. Cross reactivity of extranuclear oestrophilin with antibody to nuclear form.

oestradiol by the rat uterus to the same degree. It was found that there is a difference in saturability between the initial uptake of oestradiol by the rat uterus *in vivo* and its longer term retention by the nucleus, and that the 8 S extranuclear complex can be produced in surprisingly large amounts by adding the hormone directly to uterine cytosol. This led to the suggestion, as early as 1966, that the extranuclear 8 S protein, present in considerable reserve, might serve as an 'uptake' receptor, bringing the hormone to the nucleus where it is retained in limited amount by a nuclear receptor. In the following year it was proposed independently by Gorski and by ourselves that the nuclear receptor actually is an altered form of the cytosol receptor that has been translocated to the nucleus. At that time our hypothesis was based on three principal experimental observations: (i) no 5 S nuclear oestradiol–receptor complex is formed by treatment of immature rat uterine nuclei or nuclear extracts with oestradiol alone, but incubation of nuclei with hormone in the presence of receptor-containing cytosol gives rise to extractable 5 S complex; (ii) when excised rat uteri are exposed to oestradiol at 2 °C, cell fractionation as well as autoradiographic experiments demonstrate that most of the hormone is present as extranuclear 8 S complex, shifting to nuclear 5 S complex if the tissues are then warmed to 37 °C; and (iii) administration of oestradiol *in vivo* causes a temporary depletion of the receptor content of the uterine cytosol, consistent with its movement to the nucleus. Subsequent experimentation, discussed in the following sections, has provided additional evidence that supports the concept of the two-step receptor translocation mechanism (Table 5-1).

Receptor activation

When it was recognized that the 8 S extranuclear receptor is composed of 4 S subunits, it became evident that the alteration of oestrophilin that accompanies its hormone-induced migration to the nucleus is reflected by an increase in sedimentation rate

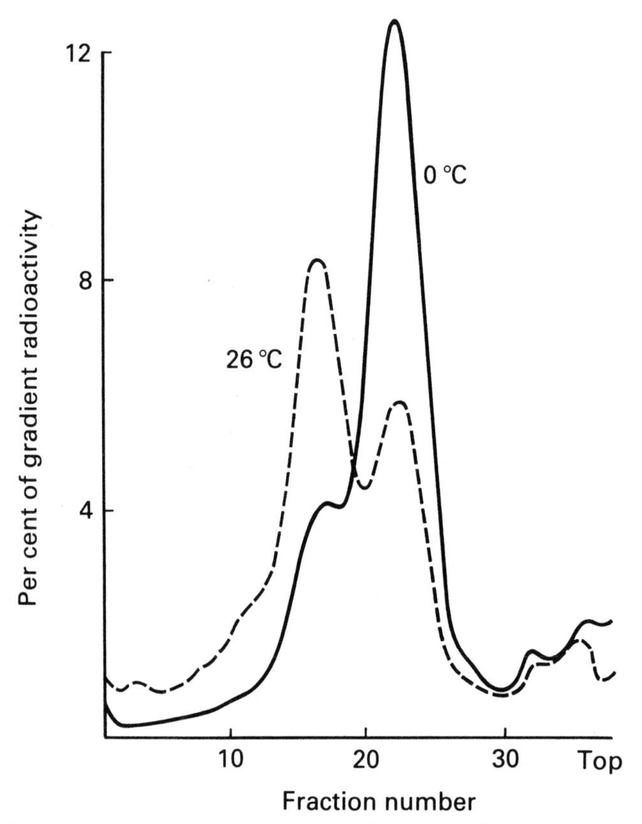

Fig. 5-11. Oestradiol-induced transformation of extranuclear receptor to the nuclear form. The cytosol fraction of a homogenate of immature rat uteri in 9 volumes of 10 mM Tris buffer, pH 7.5, was made 2 nM in [^{3}H]oestradiol and incubated for 30 minutes at either 26 °C or 0 °C, after which aliquot portions were subjected to ultracentrifugation in 5–20 per cent sucrose gradients containing 400 mM KCl. On longer incubation (60 minutes) at 26 °C, the 4 S complex (right peak) is completely converted to the 5 S form (left peak). (From E. V Jensen *et al. Biochem. Soc. Symp.* **32**, 133 (1971).)

of the hormone-binding unit from 4 S to 5 S. Originally this transformation was believed to occur in the nucleus itself, but later it was found that conversion of the 4 S cytosol complex to the nuclear form is effected simply by warming uterine cytosol to 25–37 °C in the presence of hormone (Fig. 5-11). The 5 S oestradiol–receptor complex thus produced, like that extracted from uterine nuclei, has two properties not shown by the native form; it can bind to isolated nuclei or chromatin and, as described later, it can enhance the RNA polymerase activity of isolated nuclei from hormone-dependent tissues and tumours. Because of these new properties acquired, the hormone-induced, temperature-dependent transformation of the native receptor to the biochemically functional, nuclear form is known as receptor activation.

Although the molecular details of receptor activation are not completely understood, the reaction has been shown to follow second order kinetics suggestive of dimerization. How such a process endows the activated receptor with the ability to bind to chromatin, as well as to DNA and other polyanions, is not completely clear, nor is the exact nature of the acceptor site to which the activated hormone–receptor complex binds in the chromatin. In the case of the progesterone receptor of chick oviduct (Chapter 6), it has been found that the receptor protein consists of two different components, one that binds non-specifically to DNA and the other that shows specific affinity for the non-histone proteins of target cell chromatin. It is proposed, but not established with certainty, that this dimeric molecule exerts a concerted interaction with the chromatin, resulting in the uncovering of template sites for RNA synthesis. Extension of this dual interaction model to the case of oestrogens must await experimental verification.

The fact that hormone-induced receptor activation can take place in the absence of nuclei does not necessarily mean that, in the living cell, this process may not occur preferentially in the nucleus. There are a number of experimental observations that suggest that oestrophilin sometimes is present in the nucleus uncomplexed with hormone. In contrast to results with uterus, nuclei of rat pituitary tumour and MCF-7 cell lines from human breast cancer, as well as chick and toad liver, have been found to contain substantial amounts of unoccupied oestrogen receptor even before exposure of the tissue to hormone. Similarly, it is the nucleus and not the cytoplasm of rat vaginal epithelium cells that contains a specific receptor for 5-androstene-3β,17β-diol, a C-19 steroid that elicits an oestrogen-like effect in this tissue. These findings suggest that, at least in some cases, the native form of the receptor may be distributed throughout the target cell, even though firm binding to chromatin and eventual biochemical action may require hormone-induced activation of the receptor. As discussed in a later section, more definite information about the intracellular localization of oestrophilin before exposure to hormone could be provided by techniques for identifying the receptor that do not depend on its binding to labelled hormone as a marker.

OESTROGEN–RECEPTOR COMPLEXES AND RNA SYNTHESIS

For all classes of steroid hormones, an early response is the enhancement of RNA synthesis in target cells. In the case of the primary stimulation of a hormone-deprived tissue, such as the oestrogen-induced growth of the immature rat uterus, incorporation of labelled precursors into all types of RNA is rapidly accelerated. On the other hand, in the secondary stimulation of a previously developed tissue, such as the oestrogen-pretreated chick oviduct or frog liver, a predominant effect of administered hormone is to enhance production of messenger RNAs to code for specific proteins that will be exported. A detailed account of the effect of steroid hormones on transcription is beyond the scope of this chapter. We consider here the evidence for the participation of oestrogen–receptor complexes in the tissue-specific stimulation of RNA synthesis, and the importance of receptor activation in this phenomenon.

Although the complete action of hormone–receptor complexes in target cell nuclei may involve various aspects of RNA synthesis and processing, a particularly striking effect is seen on RNA polymerase systems. Administration of oestradiol to immature rats leads to a doubling of the ability of their uterine nuclei to incorporate labelled precursors into RNA *in vitro*, whereas there is no effect on the RNA polymerase activity of liver nuclei. The template function of uterine chromatin from oestrogen-treated rats or rabbits, or of oviduct chromatin from oestrogen-treated chicks, is increased over that of corresponding chromatins from untreated animals. After oestradiol injection, both RNA polymerases I and II are stimulated in rat uterine nuclei but with different time patterns. Polymerase II activity shows an increase at 15–30 minutes, and then subsides to be followed by a second rise after 2–3 hours, whereas polymerase I activity, as well as template function of the uterine chromatin, shows a prolonged enhancement first detectable at about 1 hour. The transient early increase in polymerase II activity is not in itself sufficient for hormonal response, because a single injection of oestriol, which induces the transient stimulation of polymerase II but not the prolonged effect on both polymerase I and polymerase II, does not promote appreciable uterine growth.

The participation of hormone–receptor complexes in the enhancement of RNA synthesis in target cell nuclei is substantiated by their direct effect on the RNA polymerase activity of isolated nuclei or on the template function of target cell chromatin. The RNA polymerase I activity of uterine nuclei, while not sensitive to oestradiol itself, is doubled after the nuclei are incubated with oestradiol in the presence of uterine cytosol containing the receptor. Only the activated or nuclear form of the oestrogen–receptor complex is effective in stimulating RNA synthesis in isolated uterine nuclei, and the effect is specific for nuclei of hormone-dependent tissues (Fig. 5-12). Nuclei of hormone-dependent rat mammary tumours resemble uterine nuclei in the sensitivity of their RNA polymerase system to enhancement by oestrogen–receptor complex *in vitro*, whereas nuclei from autonomous tumours are not susceptible to such stimulation. Using either bacterial or endometrial RNA polymerase enzyme, the template function of chromatin isolated from target, but not from non-target cells is significantly increased after exposure to oestradiol–receptor complex *in vitro*.

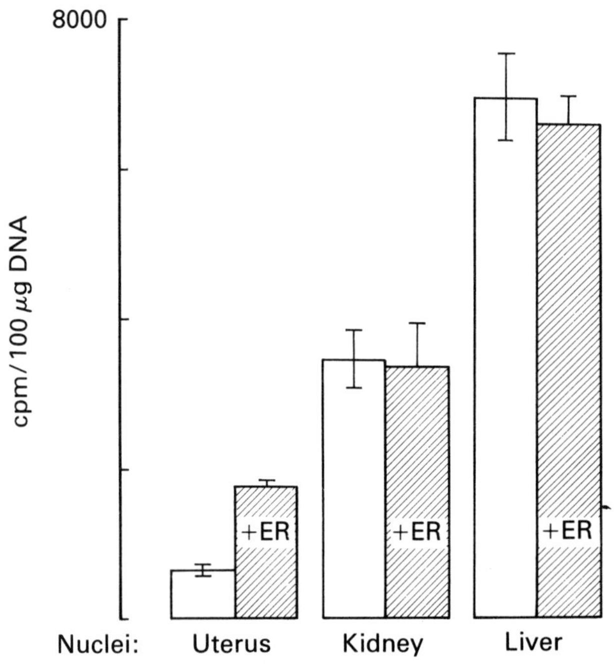

Fig. 5-12. Tissue specific influence of oestrogen-receptor complex on RNA polymerase activity of isolated uterine nuclei. Nuclei, isolated from 2.2 M sucrose homogenates of various immature rat tissues, were incubated at 25 °C for 30 minutes with rat uterine cytosol in the presence (ER) and absence of 10 nM oestradiol. The nuclei were separated by centrifugation and resuspended in 0.32 M sucrose from assay of Mg^{2+}-dependent RNA polymerase by measuring the incorporation of [^{3}H]UMP from UTP. (From E. V. Jensen *et al. J. Steroid Biochem.* **3**, 445 (1972).)

From the foregoing observations it appears that oestrogen-dependent tissues and tumours have a characteristic limitation in the activity of their RNA synthesizing or processing systems, probably involving, at least in part, a restriction on chromatin template function, which can be alleviated by an activated hormone–receptor complex of extranuclear origin. In the case of uterine growth, this action appears to require the continued presence of oestrogen–receptor complex in the nucleus for a period of several hours. It has been variously estimated that a single physiological dose of oestradiol results in the translocation of between 6 000 and 14 000 oestrogen–receptor complexes to the uterine nucleus. Thus it appears that the uterotrophic effect of oestrogen involves a process requiring the ongoing participation of substantial numbers of hormone–receptor complexes rather than the triggering of an initial event at a few specific gene sites.

ACTION OF ANTI-OESTROGENS

Because oestrogen antagonists of the clomiphene type (Fig. 5-7) compete with active oestrogens for binding to oestrophilin, their ability to inhibit the uterotrophic action of oestradiol was first believed to result simply from their preventing the hormone from binding to the extranuclear receptor and inducing its translocation to the nucleus. More recently, it has been demonstrated that the anti-oestrogens form complexes with extranuclear receptors that actually are translocated to the nucleus. Unlike the case with oestradiol, where depleted receptor is replenished (Fig. 5-13), depletion of extranuclear oestrophilin with nafoxidine is not followed by receptor resynthesis, so that the continued availability of receptor for translocation to the nucleus, believed necessary for full oestrogenic response, is not maintained. Thus, nafoxidine can act as a weak oestrogen, but by eliminating the cytosol receptor it can inhibit the uterotrophic action of a true oestrogen.

NON-GENOMIC ACTIONS OF OESTROGENS

In addition to effects resulting from the influence of oestrogen–receptor complexes on RNA production, there are certain actions of oestrogens that do not appear to involve interaction with the nucleus. It has long been known that a very rapid uterine response to oestrogen administration is intense hyperaemia leading to the imbibition of fluid. Early work by Shelesnyak and by Szego has demonstrated a role for histamine in this process, and more recently Tchernitchin has presented

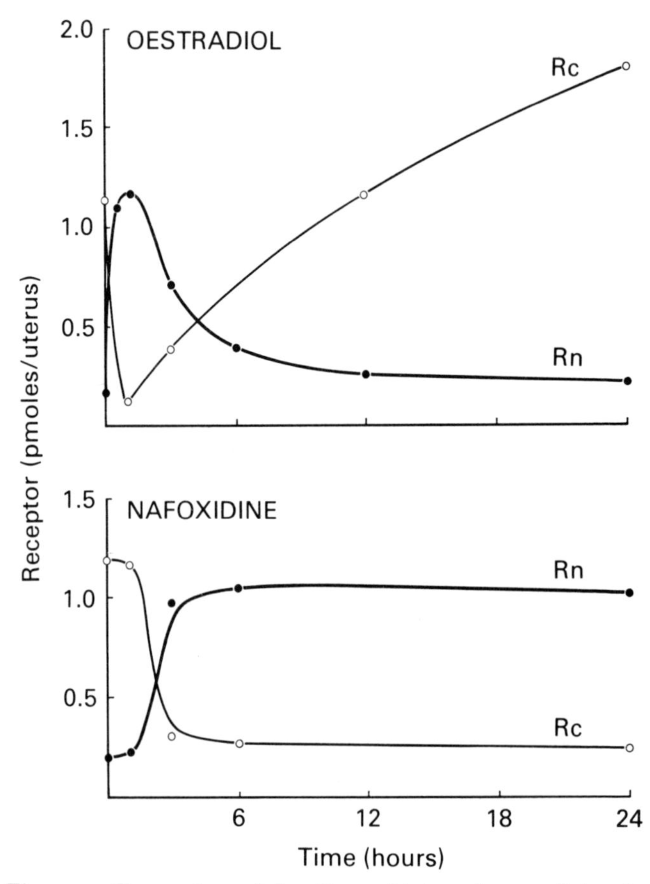

Fig. 5-13. Comparison of the effects of injected oestradiol and nafoxidine on nuclear (Rn) and cytoplasmic (Rc) oestrophilin levels in the immature rat uterus. (From J. H. Clark *et al.* In *Receptors and Hormone Action*, vol. 2, Chap. 1. Ed. B. W. O'Malley and L. Birnbaumer. New York; Academic Press (1978).)

evidence that eosinophils in the uterine circulation contain oestrogen-binding components that appear to be involved in hormone-induced histamine release. These phenomena suggest that oestrogens, and possibly other steroid hormones, may have more than one type of interaction in target cells, and that not all biochemical responses necessarily are mediated through the action of receptor complexes in the genome.

ANTIBODIES TO OESTROPHILIN

As mentioned earlier, essentially all our information about the occurrence, properties and interactions of oestrogen receptors has been obtained from experiments using the radioactively-labelled hormone as a marker for the receptor to which it binds. The recent preparation of antibodies to purified oestrophilin provides for the first time a means of detecting the receptor in the absence of bound steroid.

Immunoglobulin from the serum of rabbits, immunized with highly purified preparations of the nuclear form of the oestradiol–receptor complex of calf uterus, reacts with oestradiol–receptor complexes to yield non-precipitating products without release of the bound hormone. Thus, in addition to using conventional techniques of immunochemistry, one can conveniently study the interaction of antibody with oestrophilin by examining its effect on the sedimentation properties of the receptor, using the radioactive steroid as a marker. The antibody reacts with both nuclear and extranuclear oestrophilin, not only of calf uterus from which the immunogen was obtained, but also from oestrogen-dependent tissues of all species so far tested, including rat, mouse, guinea pig, rabbit, sheep, and monkey uterus; monkey and hen oviduct; rat mammary endometrial and pituitary tumours; and human breast cancer.

Nuclear and extranuclear forms of oestrophilin differ in their reaction pattern with the antibody. The purified calf nuclear oestradiol–receptor complex forms a single, rapidly sedimenting product (11–12 S), while crude nuclear complexes from various species yield this entity and an apparently smaller (8 S) product (Fig. 5-14). In contrast the extranuclear oestradiol–receptor complexes show only the slower peak, which sediments at 7.5 S in salt-containing sucrose gradients (Fig. 5-15). After hormone-induced activation (Fig. 5-11), the cytosol receptor

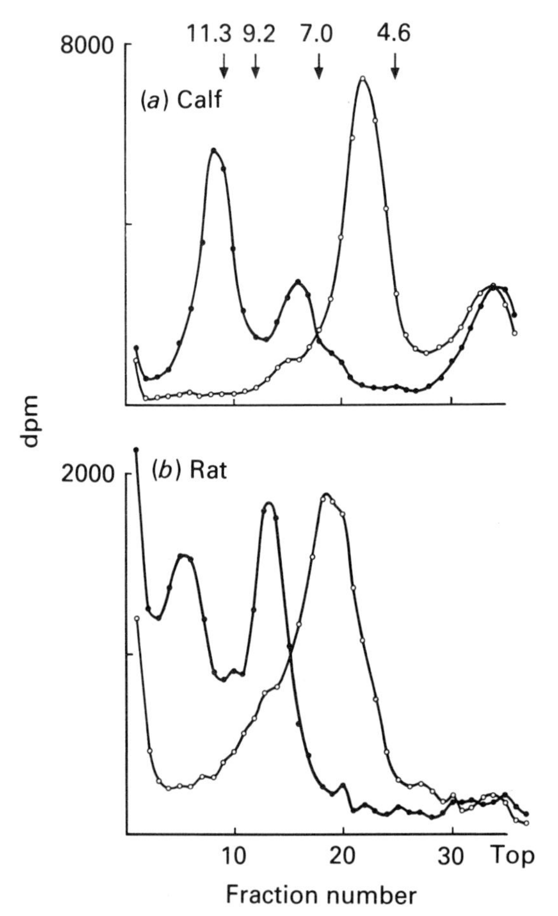

Fig. 5-14. Sedimentation pattern in 10 to 30 per cent sucrose gradients containing 400 mM KCl of: (*a*) 400 mM KCl extract of calf uterine nuclei after incubation for 60 minutes at 25 °C with 20 nM [^{3}H]oestradiol in calf uterine cytosol, and (*b*) similar extract of uterine nuclei from immature rats four hours after injection of 100 ng [^{3}H]-oestradiol *in vivo*, in the presence of immunoglobulin from immunized (●) or non-immunized (○) rabbits. (From G. L. Greene *et al. Proc. Natl. Acad. Sci. U.S.A.* **74**, 3681 (1977).)

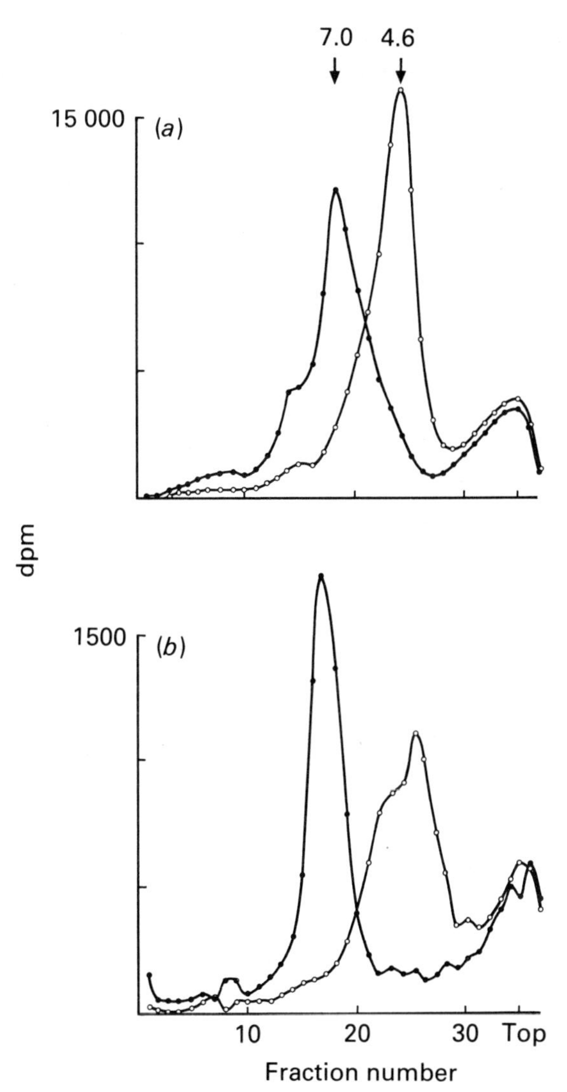

Fig. 5-15. Sedimentation pattern in 10–30 per cent sucrose gradients containing 400 mM KCl, of cytosols from: (*a*) calf uterus and (*b*) human breast cancer, each containing [^{3}H]oestradiol, in the presence of immunoglobulin from immunized (●) or non-immunized (○) rabbits (From G. L. Greene *et al. Proc. Natl. Acad. Sci. U.S.A.* **74**, 3681 (1977).)

complex resembles the nuclear one in showing both the 8 S and 11–12 S peaks in the presence of antibody.

The fact that antibodies raised against the nuclear form of oestrophilin cross-react with extranuclear as well as nuclear oestrogen receptors from many tissues provides additional evidence for the concept that the nuclear receptor is derived from the translocation of the cytosol receptor (Table 5-1). The formation of a more rapidly sedimenting immune complex with activated or nuclear oestrophilin suggests that this modification can bind more antibody units per molecule that the native form, consistent with the proposal that receptor activation may involve dimerization of native oestrophilin.

In contrast to their reactivity with oestrogen–receptor complexes from various sources, antibodies to oestrophilin do not cross react with either androgen–receptor complexes of rat prostate or with progesterone–receptor complexes from rabbit uterus, rat endometrial tumour or chick oviduct. Thus there is immunochemical similarity among oestrophilins from different tissues of a wide variety of mammalian species, but receptors for different classes of sex hormones appear to be immunologically distinct, at least in regard to the determinants recognized by this antibody. These cross-reacting oestrophilin-specific antibodies offer promise as useful reagents for the purification and assay of oestrophilin through immunochemical techniques, and may provide insight into many of the unresolved questions of receptor synthesis, intracellular localization and activation.

Progress in any field is facilitated by the ability of investigators to pose proper questions and so design informative experiments. In the case of mechanisms of steroid hormone action, recognition that the intracellular agent that delivers the regulatory signal is a hormone-activated steroid–protein complex, rather than the steroid itself, has permitted more rational utilization of broken cell systems to seek biochemical information of relevance to actual physiological processes. Until now our studies of hormone–receptor interaction have depended on the radioactive steroid to guide us to the receptor protein. This approach, though highly productive, has certain limitations in that introduction of the hormone causes perturbations in receptor distribution and the steroid may not recognize the receptor either at early stages of its synthesis (pro-oestrophilin?) or at late stages of its actions when the steroid and protein part company. Antibody that recognizes the receptor independently of its hormone binding should prove a valuable complement to the steroid as a marker for the receptor. Insights obtained from the application of immunochemical techniques to receptor studies, taken with information derived from the exciting recent advances in the structure and function of eukaryotic genes, may bring us closer to a complete understanding of the biochemical mechanism of steroid hormone action.

SUGGESTED FURTHER READING

Estrogen–receptor interaction. E. V. Jensen and E. R. DeSombre. *Science* **182**, 126–134 (1973).

Female steroid hormones and target cell nuclei. B. W. O'Malley and A. R. Means. *Science* **183**, 610–620 (1974).

The role of estrophilin in estrogen action. E. V. Jensen, S. Mohla, T. A. Gorell and E. R. DeSombre. *Vitamins and Hormones* **32**, 89–127 (1974).

Steroid–Cell Interactions. R. J. B. King and W. I. P. Mainwaring. Baltimore; University Park Press (1974).

Current models of steroid hormone action: A critique. J. Gorski and F. Gannon.: *Annual Review of Physiology* **38**, 425–450 (1976).

Steroid receptors: elements for modulation of eukaryotic transcription. K. R. Yamamoto and B. M. Alberts. *Annual Review of Biochemistry* **45**, 721–746 (1976).

The receptors of steroid hormones. B. W. O'Malley and W. T. Schrader. *Scientific American* **234**, 32–43 (1976).

Receptor proteins: past, present and future. E. V. Jensen. *Research on Steroids* **7**, 1–36 (1977).

Antibodies to estrogen receptor: Immunochemical similarity of estrophilin from various mammalian species. G. L. Greene, L. E. Closs, H. Fleming, E. R. DeSombre and E. V. Jensen. *Proceedings of the National Academy of Sciences U.S.A.* **74**, 3681–3685 (1977).

The oestrogens

Hormone–receptor interaction in the mechanism of reproductive hormone action. E. V. Jensen, K. J. Catt, J. Gorski and H. G. Williams-Ashman. In *Frontiers in Reproduction and Fertility Control*, pp. 245–262. Ed. R. O. Greep and M. A. Koblinsky. Cambridge; M.I.T. Press (1977).

Receptors and Hormone Action, vol. 2. Ed. B. W. O'Malley and L. Birnbaumer. New York; Academic Press (1978).

6 Progesterone

R. B. Heap and A. P. F. Flint

The history of progesterone dates from as far back as 1903 and a simple experiment reported by Ludwig Fraenkel, a young gynaecologist in Breslau. He removed the corpora lutea from rabbits a few days after mating, when the eggs were still in the Fallopian tubes, and found that the treatment prevented pregnancy. This was the first evidence that the ovaries contributed anything to pregnancy other than eggs. Later experiments showed that the corpus luteum was an endocrine organ, and in 1934 several research groups in the USA and Europe succeeded in isolating and identifying the hormone. The name 'progesterone' was coined by Willard Allen of Rochester, N.Y., and agreed upon at a meeting of some of the main investigators in London the following year.

BIOLOGICAL PROPERTIES

Progesterone itself is not unique to mammals, but can also be found in lower animals and in plants. In mammals it is the hormone of pregnancy. It regulates the rate of egg transport through the Fallopian tube, prepares the uterus to receive the implanting blastocyst by stimulating endometrial proliferation, inhibits ovulation by blocking the release of pituitary gonadotrophins, maintains pregnancy, and increases lobular–alveolar growth of the mammary gland. The disappearance of progesterone from the circulation at the end of pregnancy may regulate the onset of parturition and lactation in some species, and at the end of the menstrual cycle its withdrawal evidently precipitates the return of menses. In many of these activities progesterone acts in combination with other hormones, its interactions with oestrogen being especially important, producing specific effects that we shall look at more closely later. At this stage we should note that the action of progesterone frequently depends on the prior sensitization of target tissues by oestrogens, and this has often complicated the interpretation of the precise mechanism by which progestagens act.

Progesterone has other diverse actions. It, or probably its metabolites, are thermogenic in women, causing a rise in body temperature of 0.5–1.0 °F. This effect is the basis of the basal body temperature test in women for the detection of ovulation (see Book 3, Chapter 3). There have been reports, too, that progesterone promotes water excretion and influences electrolyte balance. It is also hypnotic, and in high doses some of its derivatives can be used as anaesthetics.

CHEMISTRY

Progesterone is a 21-carbon steroid with a double bond between C-4 and C-5, (identified as Δ^4), and carbonyl groups at C-3 and C-20 (Fig. 6-1). The so-called 'conjugated olefinic bond system', which includes the oxygen atom at C-3 and the Δ^4 bond endows the compound with a distinctive absorption spectrum in the ultraviolet with a maximum at 240 nm. This property was exploited in the early determinations of progesterone in biological samples. However the extinction coefficient of progesterone in ethanol ($E_{\text{molar}} = 17000$), although relatively high, restricts the sensitivity of this analytical method to about 1 μg per ml of solution to be measured. If this method were to be used for the measurement of progesterone in human peripheral venous plasma in the luteal phase of the menstrual cycle, when the levels reached are only 6–12 ng/ml, one would need almost a litre of blood! – hence the importance of modern radioimmunoassay techniques capable of measuring picogram quantities in small samples.

The chemical structures of some closely related metabolites of progesterone, and of its immediate precursor pregnenolone, are also given in Fig. 6-1. The pathways of synthesis of other

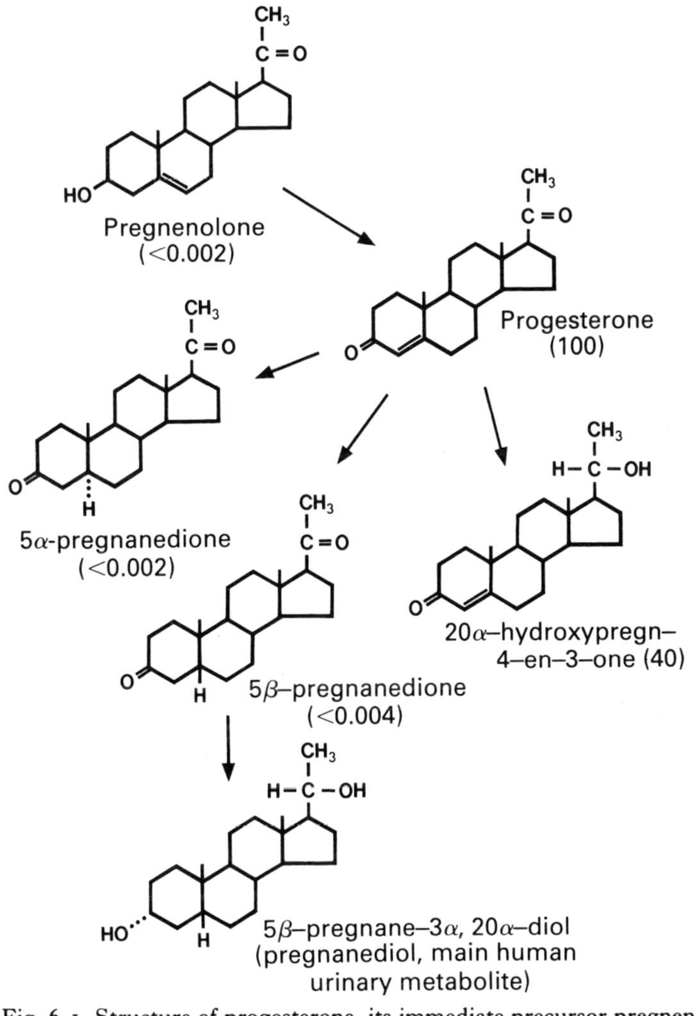

CH_3
C=O
HO
Pregnenolone
(<0.002)

CH_3
C=O
O
Progesterone
(100)

CH_3
C=O
O
H
5α-pregnanedione
(<0.002)

CH_3
H–C–OH
O
20α–hydroxypregn–
4–en–3–one (40)

CH_3
C=O
O
H
5β–pregnanedione
(<0.004)

CH_3
H–C–OH
HO
H
5β–pregnane–3α, 20α–diol
(pregnanediol, main human
urinary metabolite)

Fig. 6-1. Structure of progesterone, its immediate precursor pregnenolone, and some of its main metabolites. Values in parentheses indicate progestational activity (as a percentage of that of progesterone) in the Hooker-Forbes bioassay, which depends on the transformation of endometrial cell nuclei (becoming plump, slightly elongated in shape with conspicuous nucleoli and fine, dispersed chromatin) after intrauterine injections of small volumes (0.6 μl) of steroid in oil. Many people feel that results obtained by the Hooker–Forbes assay show suspiciously high activities for 20α-hydroxypregn-4-en-3-one and some related compounds.

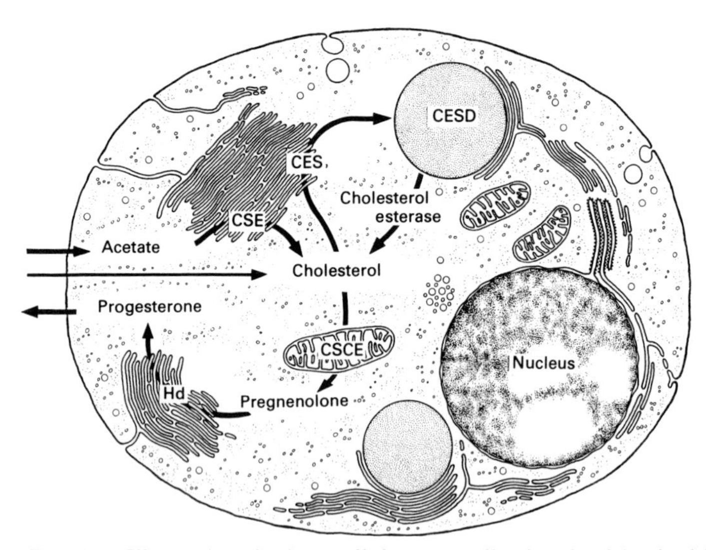

Fig. 6-2. Illustrating the intracellular organelles involved in the biosynthesis of progesterone in steroidogenic tissues. Note that in some tissues 3β-hydroxysteroid dehydrogenase is located in mitochondria as well as in the endoplasmic reticulum; that cholesterol may enter the cell from the blood; and that there may be more than one intracellular 'pool' of cholesterol. CES, cholesterol ester synthetase; CESD, cholesterol ester storage droplet; CSCE, cholesterol side-chain cleavage enzyme; CSE, cholesterol synthesizing enzymes; Hd, 3β-hydroxysteroid dehydrogenase.

steroids from progesterone, including androgens and oestrogens, have been described by David Baird in Book 3, Chapter 1. (Although it is beyond the scope of this discussion we may note in passing that corticosteroids are also synthesized from progesterone.) At least 26 metabolites may be formed from progesterone by the reduction of the C-4 double bond and the C-3 and C-20 oxo-groups, while oxidation could theoretically yield nearly 1000 metabolites. Endocrinologists can consider themselves fortunate that relatively few of these compounds are found in nature, and that even fewer are biologically active (Fig. 6-1).

Reduction of the C-4 double bond of progesterone results in the formation of 5α- and 5β-pregnanediones. The 5α- compound shows very little biological activity (Fig. 6-1), though in some test systems it can compete with progesterone for binding to its receptor in uterine cytosol. Reduction at C-20, however, results in substances that still have some biological activity.

The enzymes involved in the synthesis and metabolism of progesterone, and their intracellular localization in endocrine organs, are shown in Fig. 6-2.

BINDING PROPERTIES OF PROGESTERONE

Because of its lipoidal nature, progesterone (like other steroids) is capable of combining with many lipophilic substances. For example, it can pass into the lipid phase of cell membranes, and can bind in a non-specific way to many proteins. In some instances these properties are physiologically important.

Lipid solubility and membrane effects

The hypothesis that steroids penetrate cell membranes by orienting themselves between phospholipid molecules was advanced in 1961 by Edward Willmer in Cambridge, UK. Steroids might well form stable complexes with the membrane phospholipids, with their lipophilic ends associated with the aliphatic fatty acid chains and their hydrophilic ends orientated towards the aqueous environment. Some degree of specificity may be expected to arise during this process, since, for example, steroids with the non-planar 5β- configuration, will intercalate between other molecules less easily than the more planar 5α- reduced compounds. By measuring the rate of release of radioactively labelled ions (such as $^{42}K^+$) from the artificial vesicles known as liposomes, Willmer's hypothesis can be tested *in vitro*, and in this way progesterone can be shown to have a labilizing, rather than stabilizing effect, rendering the liposomes more unstable and more permeable to K^+.

One application of the effects of steroids on cell membranes is in the field of anaesthetics. Although progesterone itself is not sufficiently active to be a useful anaesthetic, it will cause anaesthesia if given in large enough doses; the median anaesthetic dose in mice is about 100 mg/kg administered intravenously. Some progesterone derivatives are more active: 3α-hydroxy-5α-pregnane-11,20-dione is marketed as an anaesthetic by Glaxo Ltd under the trade name Althesin. It will produce deep anaesthesia in 30–80 seconds and is widely used where short-term anaesthesia and rapid recovery is required, as for instance in dental surgery. The anaesthetic properties of progesterone derivatives are also exploited in the defence mechanisms of some water beetles, which when attacked by fish, release a milky, steroid-containing fluid from their prothoracic glands in quantities sufficient to produce temporary paralysis of their attackers. In the case of the German water beetle *Acilius sulcatus*, the secretion contains hydroxylated and unsaturated progesterone derivatives which cause reversible paralysis in goldfish at concentrations of 2–10 μg/ml.

In addition to the dramatic anaesthetic effects of high doses of progesterone derivatives, there are also more subtle effects associated with changes in circulating concentrations within the physiological range. There are, for example, many documented effects of progesterone on the electrical activity of various parts of the brain, and the reduction of electrical activity in the hypothalamus may be responsible for the inhibition of the 'Ferguson reflex'. This reflex involves the release of oxytocin in response to either genital or nipple stimulation (see Book 3, Chapter 2), and is blocked during pregnancy when progesterone levels are high. These effects are probably mediated by the direct action of progesterone or its metabolites on cell membranes.

Another example of the action of progesterone at a membrane comes from an amphibian, *Xenopus laevis*, in which the maturation of the egg beyond arrest at the prophase of the first meiotic division is dependent upon progesterone produced in the ovary by the follicular cells. Etienne Baulieu has recently provided persuasive evidence that this process involves an action of the steroid at the oocyte's plasma membrane: progesterone is still active after binding to a macromolecule (or to a bead) which cannot enter the cell, but not active when injected directly into it. An interesting aspect of this effect is that it is not specific for progesterone and many other steroids, including both 5α- and 5β-pregnane derivatives, are active.

Doubts surround the mechanism of action of progesterone in the inhibition of myometrial activity that occurs during pregnancy. In the rabbit, rat, mouse, sheep and goat, progesterone dramatically inhibits myometrial contractions. In the guinea pig it has no effect. Thus we cannot make generalizations about the exact role of progesterone in the maintenance of pregnancy. Electrophysiological studies have demonstrated that progesterone may affect myometrial contractility by alterations in membrane potential; the resting potential of the myometrial cell rises during pregnancy, returning to non-pregnant values before parturition (values for the rat are: non-pregnant, −56 mV; mid-gestation, −78 mV; at parturition, −54 mV). Mid-gestation values can be induced in spayed rats by the administration of progesterone, and this seems likely to be a direct effect of the steroid. The fall before parturition, however, is probably caused by oestrogen.

The control of myometrial contractility in the guinea pig is a subject of considerable interest. There is certainly a humoral inhibitor of uterine contractions in guinea pigs, which recent studies suggest may be related to relaxin. Whether the effect on membrane potential is a direct one, resulting from an interaction of the steroid with cell membrane lipids, or whether progesterone binds to a cytoplasmic receptor and initiates specific protein synthesis, is unknown. The fact that administered progesterone takes several hours to reduce myometrial contractility even in animals that are responsive, and the large species variation in the response, suggest that the latter alternative may be the correct one.

A second mechanism by which progesterone controls uterine contractility is through the catecholamine-activated adenylate cyclase system. The myometrium is relaxed by β-adrenergic drugs, which by stimulating β-receptors elevate intracellular concentrations of cyclic AMP, and it seems that progesterone decreases the threshold of the cell membrane β-adrenergic receptors, and increases their number. The result is that the myometrium becomes more sensitive to the relaxing effect of β-adrenergic catecholamines.

A third effect of progesterone, which may influence uterine contractility, is its inhibition of the synthesis and release of uterine prostaglandins. Although in most species pretreatment with progesterone appears to stimulate oestrogen-induced endometrial prostaglandin synthesis, in the pregnant rat, sheep and goat, uterine prostaglandin reaction is reduced by progesterone. Prostaglandins, particularly prostaglandin $F_{2\alpha}$, inhibit the stimulatory effect of β-agonists on myometrial adenylate cyclase (see Chapter 3), thereby potentiating contractility; reducing prostaglandin concentrations allows an increase in cyclic AMP levels and prevents the conduction of action potentials by the muscle cells.

The lipid solubility of progesterone renders it theoretically capable of entering cells from the extracellular fluid by diffusion, and although the existence of specific membrane-bound transport proteins has been suggested, none has been demonstrated. Likewise the secretion of progesterone by steroidogenic cells may occur by diffusion down a concentration gradient. However, an Australian group, Geoffrey Thorburn, Brian Stacy, Robert Gemmel and Colin Nancarrow, have found by electron microscopy that cells of ovine and bovine corpora lutea contain secretory granules analogous to those of protein-secreting cells, and these granules can be shown to be involved in exocytotic fusion with the cell membrane (Fig. 6-3). We cannot be quite sure that they contain progesterone, although there is growing evidence that they do; for instance, the number of granules found inside the cell drops as its progesterone content falls at the end of the cycle, and progesterone secretion can be blocked by inhibitors of microtubule formation such as colchicine and vinblastine. The existence in steroidogenic cells of secretory

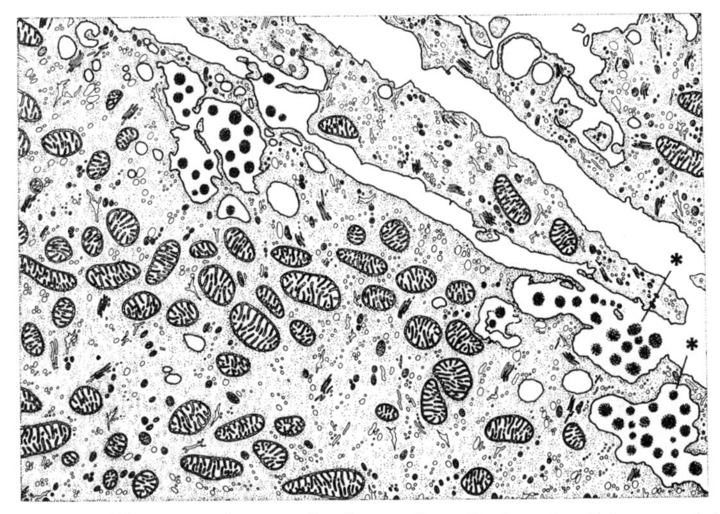

Fig. 6-3. Electron micrograph of a section of a luteal cell in material taken from a sheep on day 10 of the oestrous cycle. The dark granules (asterisked) are inferred to contain progesterone in the process of secretion from the cell. (From R. T. Gemmell, B. D. Stacy and C. D. Nancarrow. *Anat. Rec.* **189**, 161 (1977).)

granules does not, however, imply that they store progesterone: as Roger Short found some time ago, the cells of the corpus luteum contain only sufficient progesterone to maintain secretion for 10 minutes.

Binding to proteins

Steroids circulating in blood are usually bound to serum proteins (Table 6-1). Most of the blood progesterone is bound to albumin and the protein that also binds corticosteroids, identified as corticosteroid-binding globulin (CBG). It is instructive to compare these two binding mechanisms, since the carrier proteins are of different kinds. Before describing the properties of these proteins, however, we must give some thought to the quantitative aspects of protein–steroid interactions in general.

The two most common features that distinguish the way in

TABLE 6-1. Distribution of progesterone in human plasma during the first trimester of pregnancy. (From H. E. Rosenthal, W. R. Slaunwhite, Jr and A. A. Sandberg. *J. Clin. Endocr.* **29**, 352 (1969))

	Concentration (mM)	Per cent of total
Bound to corticosteroid-binding globulin	67	38
Bound to serum albumin	35	20
Bound to other proteins (including α_1-acid glycoprotein)	70	40
Unbound	4	2

which certain small molecules (called ligands) bind to proteins are the strength with which they bind (*affinity*) and the number of molecules bound per molecule of protein (*capacity*). With the aid of radio-labelled ligands and a method (such as equilibrium dialysis or addition of dextran-coated charcoal) to separate unbound from protein-bound material, the amount of ligand bound by a given amount of protein can be measured at a variety of ligand concentrations. This gives the familiar hyperbolic curve illustrated in Fig. 6-4(*a*). To determine the *affinity constant* (which is defined as the reciprocal of the free steroid concentration when the protein is half-saturated) and the *number of molecules of ligand bound*, these data are frequently plotted as a straight line by the method of Scatchard (Fig. 6-4*b*). The slope of this line is K_a ($-K_a$ = *the association constant* or *affinity constant*). The intercept on the abscissa gives the capacity, and if we know the molecular weight of the steroid and the concentration and molecular weight of the protein, we can determine the capacity in terms of mols of ligand bound per mol of protein.

The derivation of the Scatchard plot is based on the Law of Mass Action, as follows:

Assume a protein (P) binds a steroid (S) according to the equation:

$$S + P \rightleftharpoons SP.$$

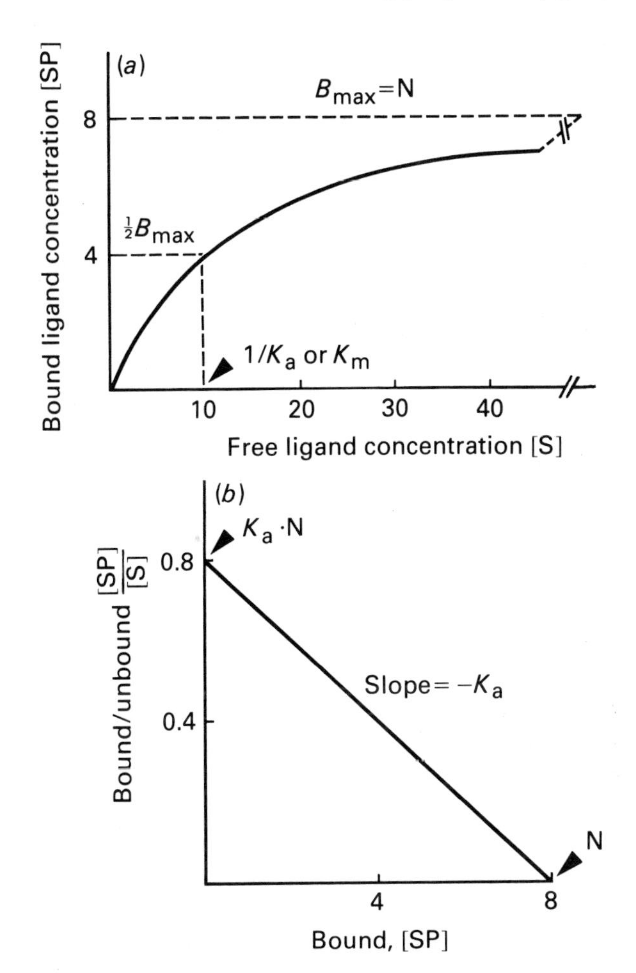

Fig. 6-4. Determination of the way in which small molecules (ligands) bind to proteins. (*a*), Michaelis–Menten plot; (*b*), Scatchard plot. K_a, association constant or affinity constant; N, total number of binding sites; P, protein; S, steroid; B_{max}, maximum binding. (After D. Rodbard. Mathematics of hormone–receptor interaction. In *Receptors for Reproductive Hormones*. Ed. B. W. O'Malley and A. R. Means. *Adv. Exptl. Med. Biol.* **36**, 289 (1973).)

At equilibrium the ratio of product to reactants is given by

$$K_a = \frac{[SP]}{[S]\,[P]}, \qquad (1)$$

where K_a is the equilibrium (or 'affinity') constant and [SP], [S] and [P] are the concentrations of steroid–protein complex, free steroid and free protein, respectively. K_a is a measure of the avidity with which the protein binds its ligand (the steroid); if the binding is very strong, [SP] (and therefore K_a) will be high. The concentrations in equation (1) are usually expressed in terms of molarity of ligand, so K_a has the dimension M^{-1} (litres/mol). When half the binding sites on the protein are occupied, that is at half saturation, [SP] = [P], and $K_a = 1/[S]$ or $[S] = 1/K_a$ (see Fig. 6-4).

Since under experimental conditions the concentration of free protein is generally not known, it is convenient to rewrite equation (1) in terms of the total number of binding sites. If their concentration equals [N] then

$$[N] = [P]+[SP]. \qquad (2)$$

Substituting equation (2) into equation (1) and rearranging we obtain

$$[SP] = \frac{[N]}{1+1/K_a \rightarrow [S]}. \qquad (3)$$

Equation (3) expresses the concentration of steroid–protein complex in terms of the total number of binding sites and the concentration of free steroid. It is analogous to the Michaelis–Menten equation used in enzyme kinetics and has the graphical form shown in Fig. 6-4(*a*). Note that binding reaches a maximum when [SP] = [N].

For the purpose of analysing experiments it has been found useful to linearize equation (3) (Fig. 6-4*b*). If we multiply both sides by $1+1/K_a \cdot [S]$ and then by K_a, after rearrangement we obtain

$$\frac{[SP]}{[S]} = K_a \cdot [N] - K_a \cdot [SP]. \qquad (4)$$

TABLE 6-2. Properties of progesterone-binding proteins in plasma

	Human serum albumin	Human corticosteroid-binding globulin	α_1-acid glycoprotein (orosomucoid)	Guinea-pig progesterone-binding globulin
		Physicochemical properties		
Molecular weight	69000	52000	41000	88000
Isoelectric point (pH)	4.9	~ 3–4	2.7	2.8
Carbohydrate content (per cent)	Nil	26.1	41.9	~ 70
		Binding properties		
K_a for progesterone* (M^{-1})	5×10^4	6×10^8	1×10^6	9×10^8
Concentration in pregnancy (mg/litre)	38000	74	750	770
($\times 10^{-7}$ M)	5500	14	180	100
Number binding sites for progesterone per molecule protein	2	1	1	1
Total capacity ($\times 10^{-7}$ M)	11000	14	180	100

* Measured at 4 °C. The binding of steroid to these proteins usually decreases at elevated temperatures (e.g. 37 °C). Measurements of affinity constants are frequently made at 4 °C.

Equation (4) is a straight line with slope K_a and $x = 0$ intercept $K_a \cdot [N]$. Since [SP]/[S] = 0 when [N] = [SP], the $y = 0$ intercept is equal to [N], the concentration of binding sites in the incubated sample of protein. If the concentration (molarity) of protein in the sample is known, [N]/[Protein] gives the number of binding sites per molecule.

Properties of some progesterone-binding proteins in plasma are compared in Table 6-2. Notice that the affinity constant of albumin for progesterone is several orders of magnitude lower than that of CBG. But since albumin is present at a much higher concentration in plasma, it binds more progesterone than one would expect on the basis of affinity constants alone. Furthermore, albumin is capable of binding 3 molecules of progesterone per protein molecule (although one site is usually occupied by a fatty acid) compared to 1 molecule bound by CBG.

Both albumin and CBG bind other steroids as avidly as they bind progesterone; in the case of CBG this binding appears to be limited to cortisol and corticosterone, but albumin binds a much wider range of steroids, as well as fats, bile acids and drugs. In all cases, the binding of these proteins appears principally to be by hydrophobic bonding; in the case of progesterone-binding globulin (PBG), it involves a tryptophan residue.

The binding of steroids by serum proteins renders them unavailable for metabolism by enzymes that normally use the free steroids as substrates. We believe therefore that steroids circulating in blood in the bound form are unavailable for uptake by target cells until they are released in the free state. This means that the effect of binding proteins is to provide a large pool of steroid that will be slowly released as the free steroid concentration falls. In other words they act as buffers against rapid fluctuations.

An example of the physiological importance of binding proteins is found in human pregnancy. Investigators had known for some time before the discovery of CBG that although there may be a four-fold rise in the circulating cortisol concentration during pregnancy, pregnant women do not have symptoms of increased adrenocorticoid secretion. Then the realization came that there was only a small rise in free cortisol during pregnancy, the major increase being in CBG-bound cortisol. The increased CBG concentration in pregnancy is due to increased hepatic synthesis under the influence of oestrogens; women on the oestrogen-containing oral contraceptive pill have a similar increase in CBG.

A parallel effect occurs in the pregnant guinea pig, where during pregnancy the circulating concentration of progesterone rises dramatically and its metabolic clearance rate drops, due to the increased production of progesterone-binding protein (PBG). The concentration of PBG in plasma rises after day-20 of pregnancy in the guinea pig. Similar plasma proteins are produced during pregnancy in other hystricomorph rodents such as the casiragua *Proechimys guairae*, and the coypu *Myocastor coypus*. In these animals, as in the guinea pig, the rise in plasma progesterone during pregnancy is due primarily to a rise in the concentration of PBGs. The synthesis of PBG in the guinea pig is not stimulated by oestrogens, but is related to the formation of the allantochorionic placenta on day 15 of pregnancy.

METABOLISM OF PROGESTERONE

Progesterone is rapidly metabolized in the body. Determination of the half-life in the circulation is generally a useful indicator of the rate at which a compound is used up; it measures the time taken for a given amount to decline to half its initial value, but for compounds like progesterone this conventional method of assessment is less informative (Fig. 6-5). The rate of removal of progesterone from blood occurs at two different rates, suggesting that the steroid is distributed interchangeably between two compartments within the body. The half-life in the inner compartment is short (about 2 minutes) and the volume of the compartment is not much greater than that of the vascular space. In the outer compartment the half-life is relatively long (15.8

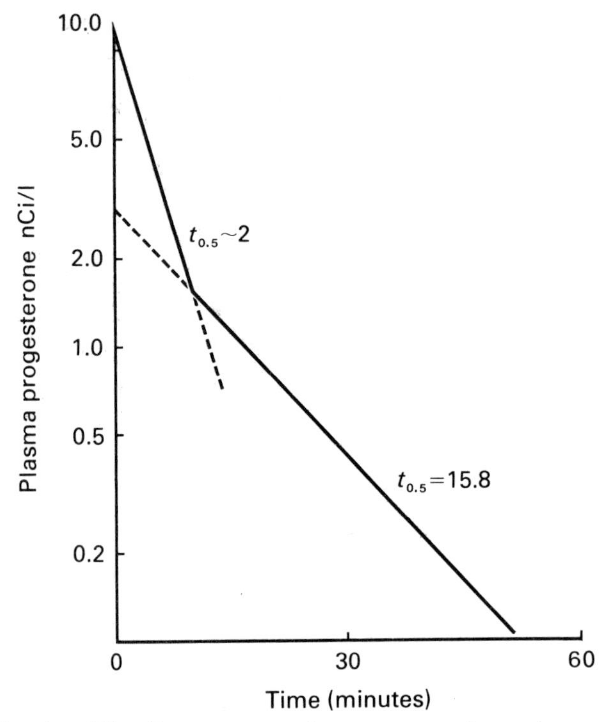

Fig. 6-5. The disappearance of progesterone from plasma after the injection of 2.0 μCi [^{3}H]progesterone into an ovariectomized woman. Results plotted on a semi-logarithmic scale show two exponentials as would be found if the steroid were distributed between two exchangeable compartments. The half-life of progesterone in the first compartment is short, whereas that in the second compartment is relatively long (15.8 minutes). (From B. Little, J. F. Tait, S. A. S. Tait and F. Erlenmeyer. *J. Clin. Invest.* **45**, 901 (1966).)

minutes) and the volume is larger. Just how rapidly progesterone is metabolized is clear from the finding that during a prolonged and constant infusion of the hormone, only a few per cent remain in blood as the original steroid, the rest being in the form of various metabolites, including the glucuronides and sulphates. The liver is a particularly active site of metabolism as it removes almost all the progesterone it receives. Although in-vitro experiments show that other organs, including target organs such as the uterus and mammary gland, are also active sites of metabolism, there is no reason to believe that a unique pattern of progesterone metabolism occurs in these tissues, nor that metabolites are formed there that are more active biologically than progesterone itself, unlike the situation with testosterone (see Chapter 4). It is true that progesterone metabolites reduced at the C-5 position (i.e. lacking $\Delta^{4,5}$) are formed in endometrial tissue incubations, comparable to the testosterone–5α-dihydrotestosterone conversion. But although 5α-dihydroprogesterone competes with progesterone for binding sites in certain target organs, it has little biological activity.

SPECIFIC BINDING IN TARGET TISSUES

The specific binding of progesterone to mammalian target tissues was first discovered by Richard Falk and Wayne Bardin working with guinea pigs (Fig. 6-6). Radioactively-labelled progesterone of high specific activity was administered to ovariectomized animals that had been primed with oestrogen (endogenous progesterone secretion had been reduced by ovariectomy and binding sites for progesterone were therefore exposed). Uterine uptake of progesterone was greater than in the heart or diaphragm, and the uterine concentration exceeded that in blood. Almost 90 per cent of the radioactivity in the uterus was in the form of labelled progesterone. Prior injection of other non-radioactive steroids failed to prevent binding of radioactive progesterone showing that uptake was largely specific. Unlabelled progesterone saturated the binding sites and radioactive uptake was reduced. A notable finding in these experiments was that pretreatment of the animals with oestrogen increased uptake of labelled progesterone. All these observations can be explained in terms of alterations in the distribution, the specificity and the control of intracellular receptors for progesterone.

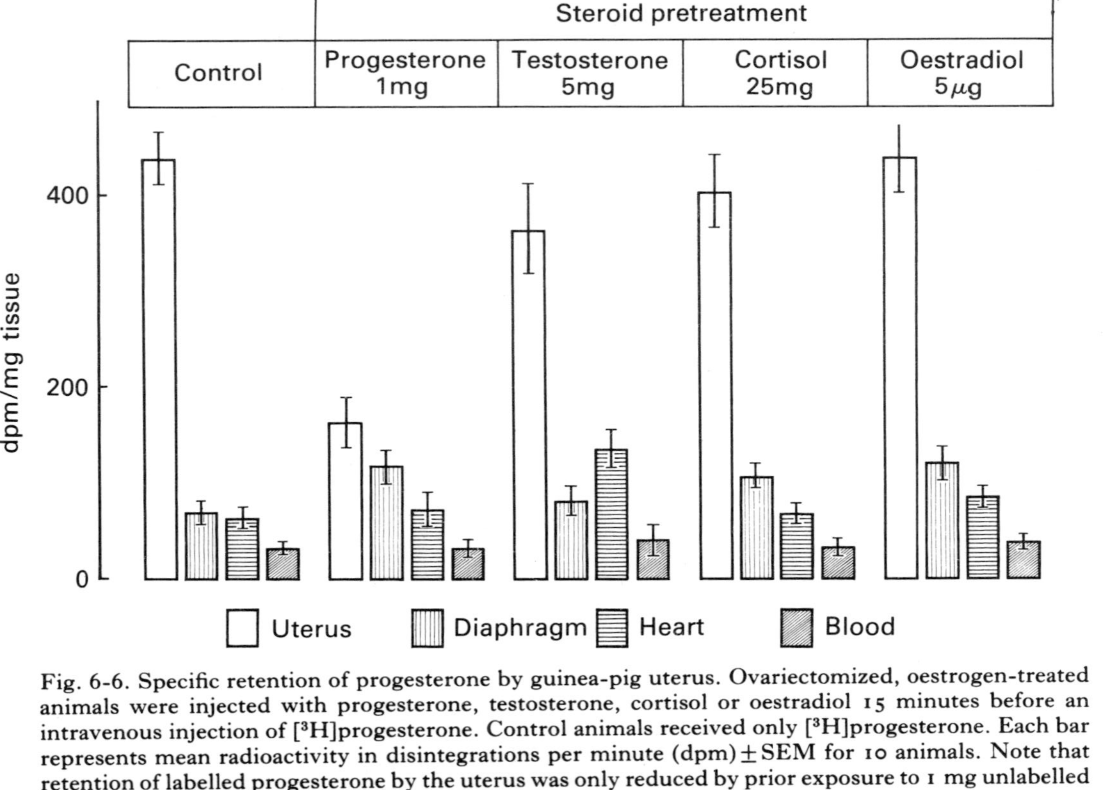

Fig. 6-6. Specific retention of progesterone by guinea-pig uterus. Ovariectomized, oestrogen-treated animals were injected with progesterone, testosterone, cortisol or oestradiol 15 minutes before an intravenous injection of [^{3}H]progesterone. Control animals received only [^{3}H]progesterone. Each bar represents mean radioactivity in disintegrations per minute (dpm) ± SEM for 10 animals. Note that retention of labelled progesterone by the uterus was only reduced by prior exposure to 1 mg unlabelled

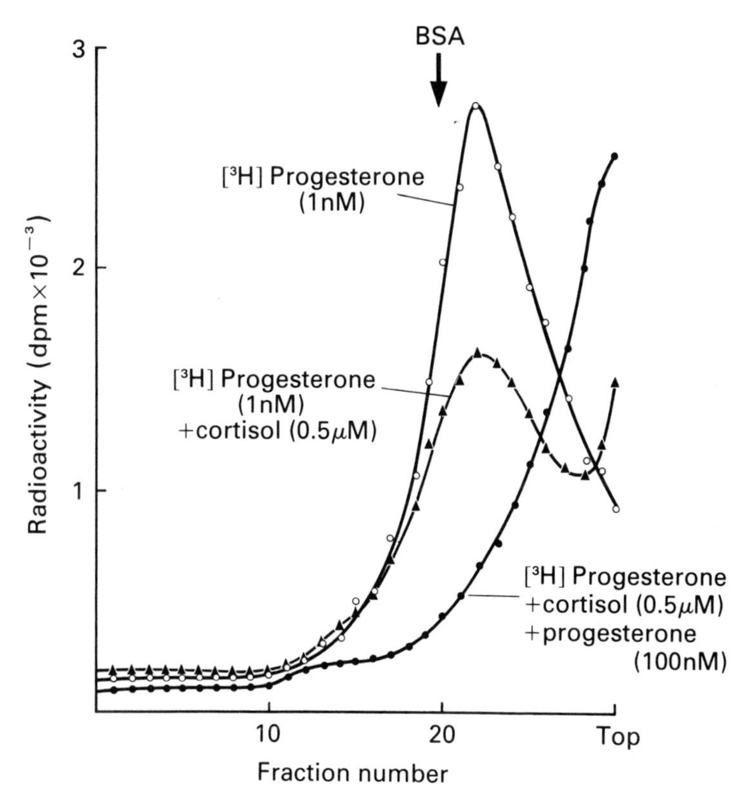

Fig. 6-7. Progesterone-binding components in human endometrial cytosol. Glycerol gradient centrifugation of 0.25 ml labelled cytosol layered on top of a linear 5–35 per cent glycerol gradient and centrifuged at 150000*g* for 18 hours at 1 °C. Bovine serum albumin (BSA) was used as a marker. Note that [^{3}H]progesterone was bound by a saturable receptor. The high binding found in fractions 18–25 was reduced by the addition of cortisol, indicating the presence of corticosteroid-binding globulin, and it was almost completely inhibited by the extra addition of progesterone, indicating the presence of a saturable receptor (4 S). (From F. Bayard, S. Damilano, R. Robel and E. E. Baulieu. *J. Clin. Endocr. Metab.* **46**, 635 (1978).

Cytosol receptors

Progesterone receptors have now been identified in several progesterone target tissues of both mammalian and non-mammalian vertebrates. They can be extracted by homogenization, and after centrifugation ($> 100000 \times g$) are found in the soluble fraction (cytosol). If the cytosol is incubated with [^{3}H]progesterone for several hours at 4 °C, the receptor protein becomes labelled, and can then be identified by centrifugation in a sucrose or glycerol density gradient. The use of density gradient centrifugation not only demonstrates the presence of a receptor protein, but also allows the determination of its sedimentation coefficient, S (Fig. 6-7). All these procedures are carried out in the cold, both to preserve the receptors and to take advantage of their increased affinity at low temperatures.

The mechanism of action of progesterone is most fully understood in the chick oviduct, largely through the work of Bert O'Malley and his colleagues in Houston, Texas. At the time O'Malley began his work, the chick oviduct was the only organ known to produce a specific, progesterone-dependent protein (avidin), which had already been characterized and could readily be estimated. Although progesterone-dependent proteins have

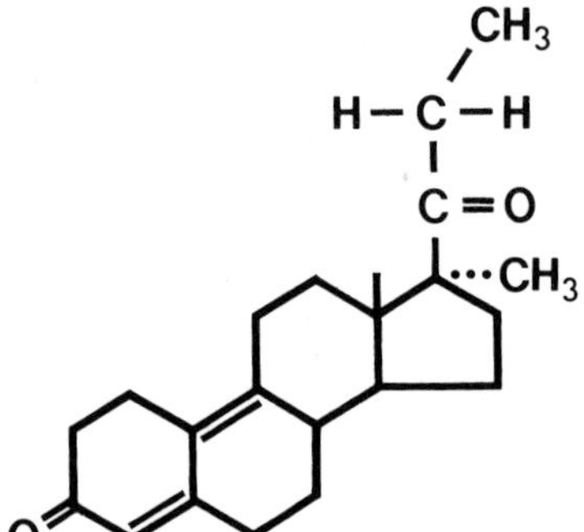

CH_3
H–C–H
C=O
CH_3
O

Fig. 6-8. Structure of the synthetic progestagen, R5020, which avidly binds to specific cytosol receptor for progesterone, but does not bind to CBG.

subsequently been studied in mammalian target organs, for a variety of technical reasons these have proved to be less amenable to investigation, so the progesterone receptor of the chick oviduct has been more thoroughly investigated than mammalian receptors.

The cytosol receptor for progesterone in the chick oviduct comprises only about 0.02 per cent of the cytosol protein, but following oestrogen treatment its concentration increases ten-fold. After chromatographic purification on diethylaminoethyl cellulose, the receptor dissociates into A and B subunits. The molecular weights of these subunits are similar and both of them possess a single high-affinity binding site for one molecule of progesterone.

A similar progesterone binding component has been found in the mammalian uterus (guinea pig, rat, rabbit, calf, hamster, sheep and man). As shown in Fig. 6-7, the presence of CBG in uterine cytosol can obscure binding by the receptor and this was a hindrance in early studies on progesterone receptor in mammalian tissues. The concentrations of the progesterone receptor and of CBG were increased by oestradiol treatment, binding was greatly reduced after heating at 60 °C for 20 minutes, and the dissociation constant was about 10^{-9} M. However, later studies showed convincingly that a cytosol progesterone receptor really does exist in the mammalian uterus. This was achieved by using a labelled synthetic progestagen (R5020, Fig. 6-8) which binds to progesterone receptors even more strongly than progesterone itself, but does not bind to CBG. So it seems that the uterus contains a high-affinity progesterone receptor in the cytosol as well as intracellular CBG which binds progesterone, cortisol and cortisone. The biological function of this intracellular CBG is not known.

Transfer of the progesterone–receptor complex to the nucleus

The next step in the action of progesterone on target cells involves the transfer of the steroid–receptor complex to the

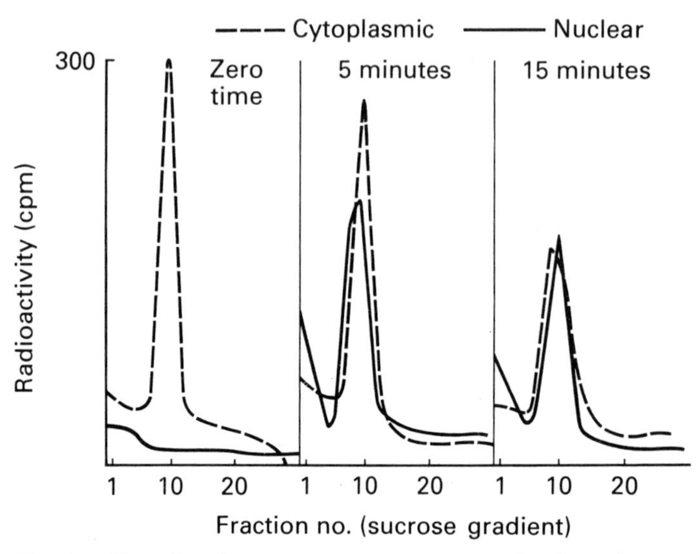

Fig. 6-9. Transfer of progesterone–receptor complex from the cytosol to the nucleus in the chick oviduct *in vitro* at 37 °C. Oviduct segments were incubated with [^{3}H]progesterone for 5 minutes at 0 °C. At zero time the steroid had formed a complex with the cytoplasmic receptor as revealed by centrifugation in a sucrose gradient. Subsequent incubations at 37 °C led to a progressive increase in the proportion of steroid found in the nucleus. (From B. W. O'Malley, M. R. Sherman and D. O. Toft. *Proc. Natl. Acad. Sci. U.S.A.* **67**, 501 (1971).)

nucleus. When immature chicks are injected with labelled progesterone, it is rapidly bound to the cytosol receptor, and subsequently the concentration of the receptor in the cytosol decreases with time, while its nuclear concentration increases (Fig. 6-9). This is consistent with the proposal of Elwood Jensen and others, from their work on the action of oestrogens, that the steroid–receptor complex is transferred from the cytoplasm to the nucleus of target cells (Chapter 5). This interpretation is also supported by the finding that the steroid–receptor complex in the nucleus is indistinguishable from that found in the cytosol in terms of its affinity constant, specificity and sedimentation coefficient.

Translocation of the receptor to the nucleus does not occur

until progesterone has bound to it as shown by the following experiments. Purified nuclear and cytosol fractions were prepared from chick oviduct and a variety of avian non-target tissues (lung, liver, spleen and erythrocytes) that do not contain the cytosol receptor. When purified oviduct nuclei were incubated with labelled progesterone little steroid was bound. But when progesterone and oviduct cytosol were added together, nuclear binding occurred. Cytosols from non-target tissues were not active, and nuclei from lung and liver failed to bind progesterone even when they were incubated with oviduct cytosol. This led to the idea that the nuclei of target cells contained *acceptor sites* with a specific affinity for the cytosol–receptor complex, and this has become known as the 'nuclear acceptor' hypothesis.

Nuclear acceptor sites and the progesterone-receptor complex

In view of the importance of the interaction between the steroid–receptor complex and the nucleus, it became essential to identify the chemical nature of the nuclear acceptor sites in order to understand the action of progesterone. Another series of ingenious cross-over experiments led to the discovery that the progesterone–receptor complex was capable of binding to specific acidic proteins in chromatin. Basic (histone) and acidic (non-histone) proteins were selectively dissociated from the chromatin of oviduct nuclei by treatment with solutions of different ionic strength and pH. Their separation from the chromatin left native DNA, to which the proteins were then added back in different combinations (Fig. 6-10). This approach was used to show that histones were not involved in binding the progesterone–receptor complex whereas the acidic proteins were, and a fraction designated acidic protein fraction 3 (AP_3) was isolated as the active component. Later AP_3 was found to bind the subunit B and the intact form of the receptor, but not subunit A. Subunit A, in contrast to the intact receptor, has a high affinity for DNA of any cell and we shall return to this point

Progesterone

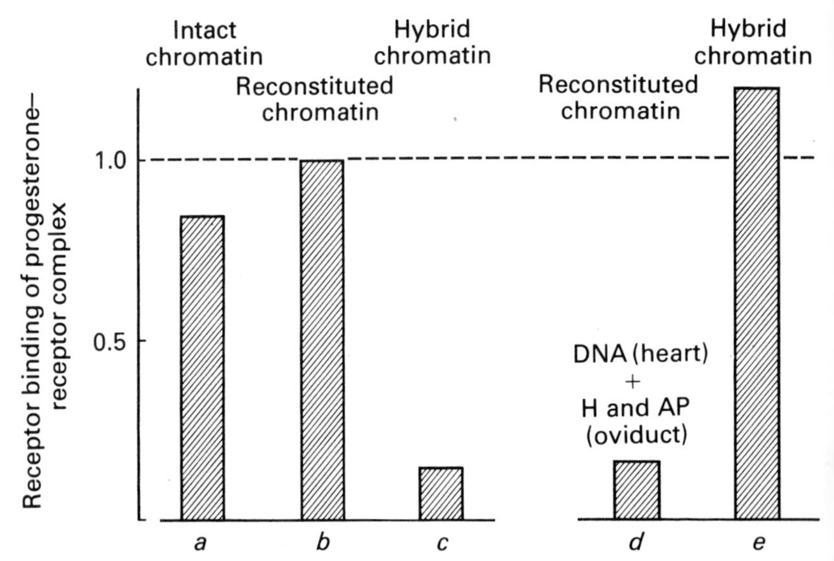

Fig. 6-10. The role of acidic non-histone proteins in the nuclear binding of the progesterone–receptor complex in target tissues. Incubations of labelled progesterone cytosol receptor showed binding to chick oviduct chromatin (*a*), reconstituted oviduct chromatin (*b*), or reconstituted hybrid chromatin containing DNA from heart muscle and histones (H) and acidic proteins (AP) from chick oviduct (*e*). Binding was low when incubated with reconstituted chromatin from heart muscle (*d*), or DNA from the chick oviduct reconstituted with histones and acidic protein from erythrocytes (*c*). (From T. C. Spelsberg, A. W. Steggles. F. Chytil and B. W. O'Malley. *J. Biol. Chem.* **247**, 1368 (1972).)

later when considering the way in which the steroid–receptor complex is able to designate appropriate genes and thus direct the synthesis of new proteins.

Masking of nuclear binding sites

Before leaving the topic of how the progesterone–receptor complex from target cell cytosol binds to specific sites in the nucleus, there is another important though controversial factor we should consider that has to do with regulating the transfer of the receptor complex to the nucleus. Thomas Spelsberg and

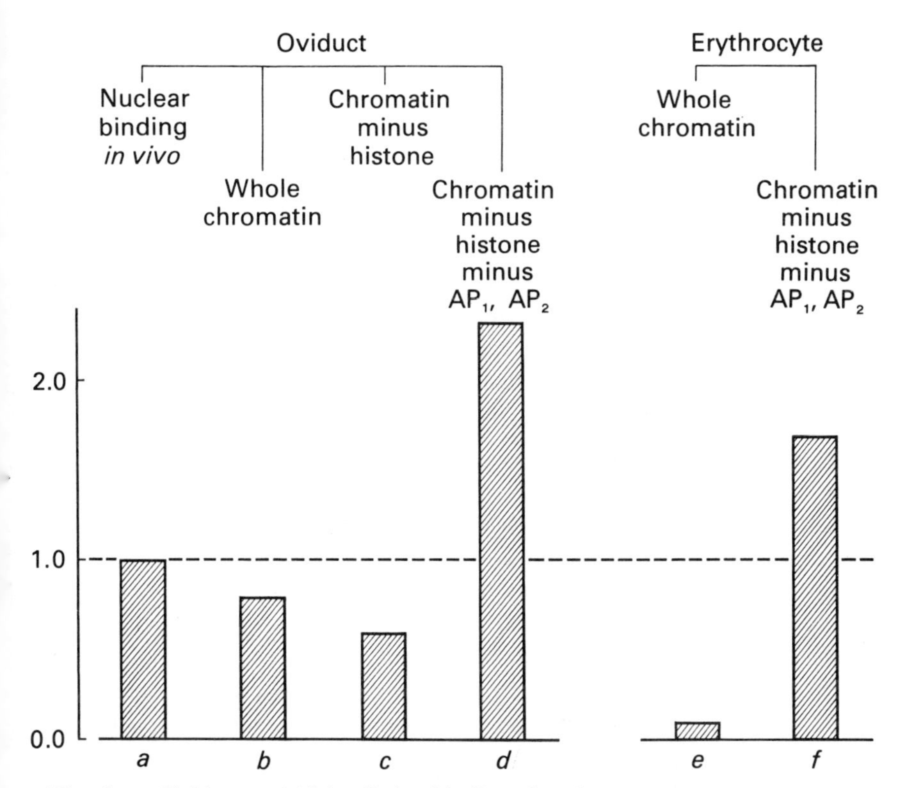

Fig. 6-11. Evidence of high affinity binding sites for progesterone–receptor complex in the nuclear material of target and non-target cells of the chicken. The results are expressed in relation to the binding of the [^{3}H]progesterone–receptor complex per oviduct nucleus *in vivo* (= 1.0). Binding to chromatin, or to fractionated chromatin, was determined *in vitro* and expressed as the number of molecules of [^{3}H]progesterone bound per cell. Removal of acidic proteins AP_1 and AP_2 from chromatin of non-target cells unmasked nuclear binding sites for the progesterone–receptor complex. (From T. C. Spelsberg, R. A. Webster and G. M. Pikler. *Nature* **262**, 65 (1976).)

his colleagues have found that nuclear binding sites for the progesterone–receptor complex also exist in the chromatin of non-target tissues (such as spleen and erythrocytes), but that they are masked by other molecules. In a target tissue such as the chick oviduct, some of these sites are uncovered. When the basic proteins and the AP_1 and AP_2 fractions of acidic protein are removed from the chromatin of the chick oviduct, the number of binding sites increases (Fig. 6-11). When the same proteins are removed from the nuclei of non-target tissues, even their binding activity is increased to almost that found in oviduct chromatin treated in a similar way (cf. Fig. 6-11*d* and *f*). But the nuclear acceptor molecules associated with AP_3 must remain with DNA, otherwise the high affinity for the steroid–receptor complex is lost. Controversy exists, however, about the identification of nuclear acceptor sites and their masking by acidic proteins. This arises from doubts about the selective fractionation of chromatin from target and non-target tissues and confirmatory studies are urgently needed in this field of investigation.

All these complex interactions take place in remarkably small dimensions. The magnitude of the molecular interactions is indicated by some figures given by Spelsberg. If one steroid–receptor complex binds to one acceptor protein, there would have to be about 20000 to 50000 acceptor molecules per oviduct cell. The molecular weight of an acceptor protein is about 15000. A somatic cell in the hen contains 2.5 pg DNA, so that there are about 10^{-15} g acceptor proteins per cell or 0.2–0.5 μg acceptor protein per mg DNA. The yield of DNA is about 1 mg/g oviduct, so that about 5 kg oviduct would be required to yield 1 mg acceptor – assuming 100 per cent recovery.

These findings show that target cells possess minute quantities of characteristic acceptor molecules which, in conjunction with DNA, bind a steroid–receptor complex with a specific configuration. This interaction of the receptor complex and the nucleus results in a unique response, which in the case of the chick oviduct is the progesterone-stimulated secretion of a new protein, avidin.

PROTEIN SYNTHESIS

The mechanism by which progesterone stimulates protein synthesis in target cells has been studied in depth in three biological systems. The hormonally-sensitized chick oviduct has proved an invaluable experimental model, and recently, uteroglobin synthesis in the rabbit uterus and purple protein production in the pig uterus have been shown to be potentially useful mammalian test systems. In contrast to its stimulatory effect on protein synthesis, progesterone can inhibit protein synthesis in certain circumstances; one example is its effect on lactose synthetase in the mammary gland where the activity of the enzyme is blocked by the presence of progesterone.

Avidin

As early as 1924, Boas noted that a diet containing 20 per cent raw egg-white produced a nutritional deficiency in rats, which results in a scaly dermatitis, general debility and eventual death. This specific toxic syndrome was caused by a factor that was named avidin because of its remarkable avidity for the B vitamin, biotin, resulting in a condition known as 'egg-white deficiency'. The biological role of avidin in the chicken's egg is unknown, but its occurrence has provided molecular biologists with an elegant model with which to study how progesterone acts.

Stimulation of synthesis. Avidin is a secretory protein of the reptilian and avian oviduct. In the chick, the stimulation of avidin synthesis can be achieved with progesterone alone; a single injection of 2 mg progesterone causes immature chicks to produce avidin 10–20 hours later. However, interactions between oestrogens and progesterone affect the rate and degree of stimulation, and synthesis is greater when both steroids are administered.

Sequential treatment of immature chicks with oestrogens and progesterone produces a characteristic pattern of tissue differ-

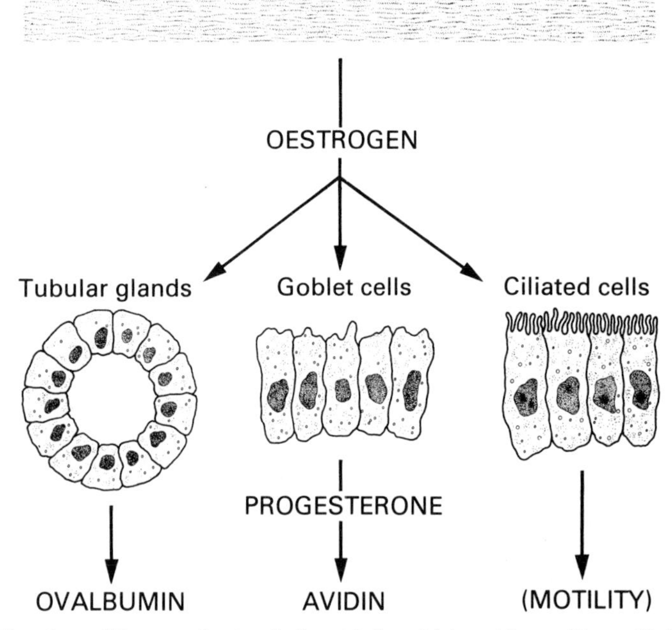

Fig. 6-12. Hormonal stimulation of the chick oviduct. (From B. W. O'Malley, W. L. McGuire, P. O. Kohler and S. G. Korenman, *Rec. Progr. Horm. Res.* **25**, 105 (1969).)

entiation. Oestrogen stimulates the oviduct to grow and secrete egg-white proteins that surround the yolk of the egg. High doses of oestrogen cause the oviductal epithelium to differentiate into tubular gland cells which secrete ovalbumin, and ciliated cells which are concerned with egg transport. After additional treatment with progesterone, goblet cells are formed which produce avidin (Fig. 6-12). This induction of avidin synthesis can only be achieved with progesterone, certain related oxo-steroids with a Δ^4-3-oxo configuration, and some synthetic progestagens.

Stimulation of avidin synthesis by progesterone does not depend on new DNA synthesis; the hormone fails to stimulate [^{3}H]thymidine incorporation into DNA and treatment with

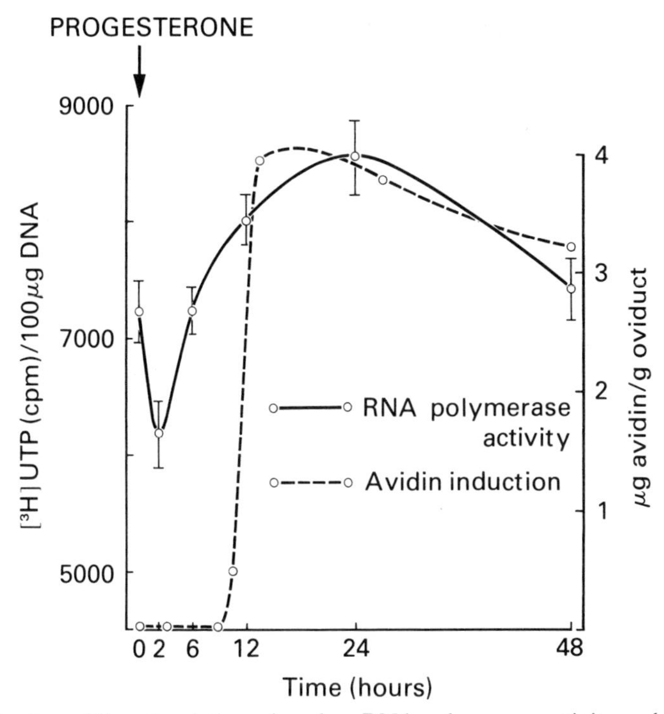

Fig. 6-13. The stimulation of nuclear RNA polymerase activity and of avidin synthesis in the oviduct of oestrogen-treated chicks after a single injection of progesterone (5 mg subcutaneous). Nuclei were isolated from the oviduct of immature chicks and RNA polymerase activity was measured as the amount of [^{3}H]uridine triphosphate, [^{3}H]UTP, incorporated into acid-insoluble product per mg of DNA in the nuclear suspension. Avidin was measured by its specific binding affinity for [^{14}C]biotin. (From B. W. O'Malley, W. L. McGuire, P. O. Kohler anbd S. G. Korenman, *Rec. Progr. Horm. Res.* **25**, 105 (1969).)

inhibitors of DNA synthesis does not interfere with avidin synthesis. Inhibitors of transcription such as actinomycin D do, however, block avidin synthesis in tissue cultures of chick oviduct, and thus it appears that the hormone exerts its effect by influencing the transcription of DNA.

RNA polymerases. These experiments with inhibitors imply that progesterone affects RNA polymerase activity, and this has been measured directly. Bert O'Malley and his colleagues found that up to 2 hours after treating oestrogen-sensitized immature chicks with progesterone, there is an initial fall in RNA polymerase activity, but this is followed by a marked increase, reaching a peak about 24 hours later. This increased RNA polymerase activity precedes the induction of avidin synthesis (Fig. 6-13).

The synthesis of new mRNA. An important piece of evidence to show that progesterone stimulates the synthesis of new protein by its action on target cell nuclei is the demonstration of new specific molecules of RNA. Activation of RNA polymerase in the oviduct of progesterone-treated chicks results in the production of a variety of new RNA molecules, including avidin messenger RNA (mRNA), for which direct evidence has now been obtained. The purification of this specific mRNA from hen oviduct is a considerable biochemical achievement, as even avidin itself is present in very low concentrations and constitutes less than 0.5 per cent of the total cellular protein. Although specific mRNA molecules had been isolated previously from eukaryotes, including those coding for albumin, ovalbumin, globin and immunoglobin, in all these cases the task was easier since the translated protein was the major product of the cell.

In the purification of specific mRNA, advantage is taken of the fact that mRNA molecules contain sequences of repeated adenine nucleotides (poly A). Avidin mRNA was prepared from an RNA fraction rich in poly A which was separated from most of the DNA and other types of RNA (transfer and ribosomal) by affinity chromatography. After this initial step, it was chromatographed a second time on a column of Sepharose 4B which removed the small molecular weight mRNAs from larger mRNAs such as that for ovalbumin. The smaller molecular weight mRNAs were then fractionated by agarose gel electrophoresis and one fraction was identified that contained avidin mRNA. Throughout these separation stages the activity of each avidin mRNA preparation was monitored by the use of an in-vitro

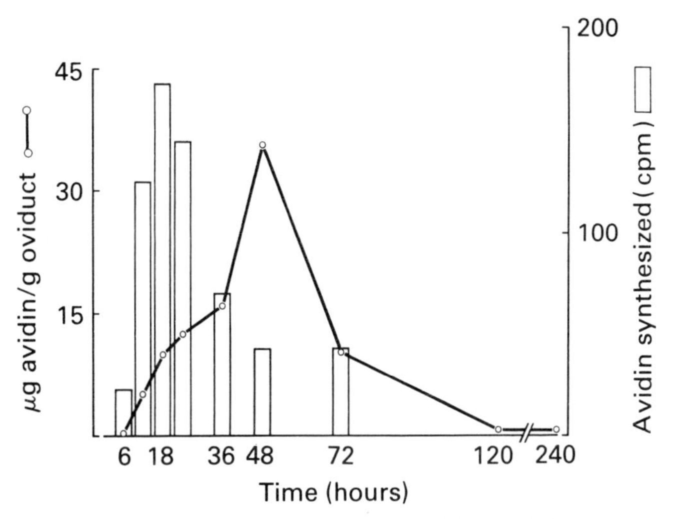

Fig. 6-14. Progesterone stimulates avidin mRNA activity (columns) before the accumulation of avidin (solid line) in the chick oviduct. Chicks treated with oestrogen for 12 days were given an injection of 1 mg progesterone at zero time. The amount of avidin mRNA in the chick oviduct was determined *in vitro* using a cell-free translation system. The concentration of avidin was measured in cytosol prepared from oviduct tissue. (L. Chan, A. R. Means and B. W. O'Malley. *Proc. Natl. Acad. Sci. U.S.A.* **70**, 1870 (1973).)

translation system. This consisted of a cell-free protein-synthesizing preparation from wheat germ which produced avidin when avidin mRNA molecules were added. Purified avidin mRNA was finally characterized by sucrose gradient centrifugation. These techniques of affinity chromatography, molecular sieving and electrophoresis resulted in a 1000-fold increase in the purity of specific avidin mRNA.

Avidin mRNA was present only in the progesterone-treated chick oviduct, and its intracellular concentration was directly dependent on progesterone stimulation. After a single injection of progesterone, avidin mRNA activity was detected within 6 hours, and maximum concentrations were maintained from 12–24 hours, thereby preceding the time when the maximum amount of avidin appeared in the tissue (Fig. 6-14). Avidin mRNA was absent from chick oviduct after oestrogen treatment, confirming the idea that progesterone specifically stimulates avidin production.

Hypothesis. Based on these and other studies, and taking into consideration the action of oestrogens on ovalbumin synthesis (Chapter 5), the hypothesis summarized in Fig. 6-15 has been proposed. The progesterone–receptor complex in chick oviduct cells, a protein composed of two different subunits A and B, is located in the cytoplasm of the target cell until the arrival of progesterone results in its translocation to the nucleus. Each subunit binds a molecule of progesterone. The A-subunit–progesterone complex is thought to regulate gene function by its binding to DNA, while the B-subunit–progesterone complex dictates the position that the A-subunit occupies on the chromatin. In this hypothesis, the A-subunit destabilizes part of the chromatin DNA, enabling RNA polymerase to attach and initiate the synthesis of new and specific mRNA. This newly-synthesized mRNA is transferred to the cytoplasm, where it serves as a template for the synthesis of proteins specific to the target cell.

A feature of the hypothesis that we have not yet considered is whether progesterone treatment increases the number of sites where the RNA polymerase enzymes can attach to the chromatin, or whether it only increases enzyme activity. Experiments in Bert O'Malley's laboratory have shown that the initiation of RNA synthesis by RNA polymerases involves the binding of these enzymes to DNA. Techniques have been developed to discover whether progesterone treatment increases the number of initiation sites where the enzyme binds to the DNA and transcription of each gene begins. Since there is only one initiation site per gene, measuring the number of these sites gives us the number of active genes. The technique involves the use of rifampicin, an antibiotic that permanently inactivates RNA polymerase molecules unless they are attached to an initiation site. The number of initiation sites is known to be increased by

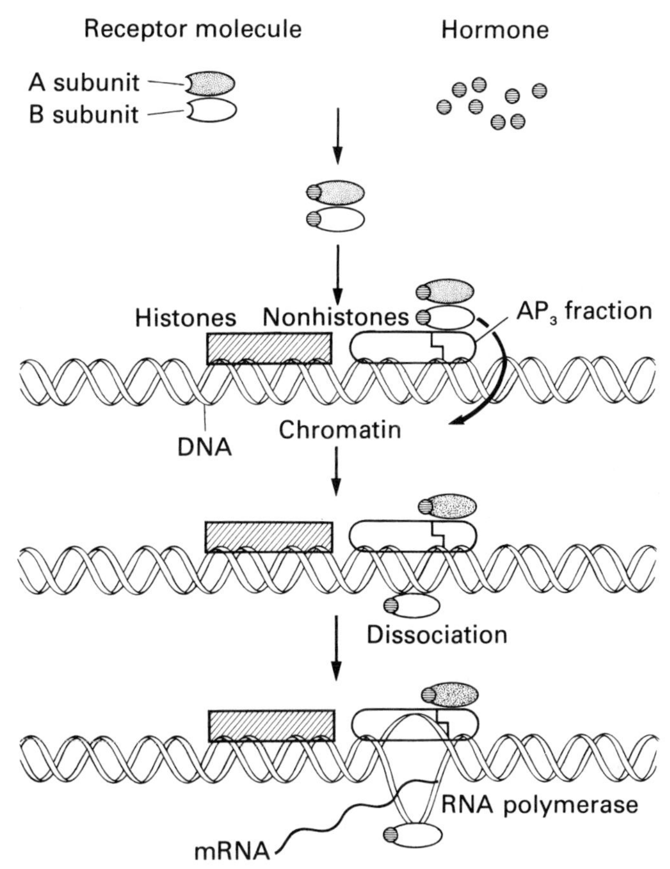

Fig. 6-15. Illustrating the hypothesis describing the mechanism of gene activation by steroid hormones in target cells. After binding 2 molecules of hormone to 1 receptor molecule, the complex is translocated to the nucleus where it binds through the B subunit to the AP_3 component of chromatin proteins. The complex then dissociates and the A subunit still bearing its hormone molecule attaches to the DNA, causes the helix to open, and allows a molecule of RNA polymerase to occupy an initiation site. This results in the synthesis of new mRNA. (From B. W. O'Malley and W. T. Schrader. *Sci. Amer.* **234**, 2 (1976).)

progesterone and to be directly related to the number of hormone–receptor complexes bound to chromatin. The number of complexes does not influence either the size of the RNA molecules formed or the rate at which they are produced. If the hormone is absent, there is no stimulation of the number of initiation sites, and if target cell chromatin is replaced by non-target cell chromatin, the number of sites is reduced.

This hypothesis is consistent with other findings on the control of protein synthesis by steroid hormones, but as yet it does not answer questions about how gene activity is switched off once initiated, or what happens to receptor molecules after they have dissociated from their nuclear binding sites. However, not all workers accept the idea that progesterone acts directly at the transcription level. There are many other points of interaction where the steroid–receptor complex may regulate protein synthesis and these include the increased release of preformed mRNA molecules from the nucleus, the transfer of mRNA into the cytoplasm, the assembly and production of ribosomal subunits and the process of mRNA translation. As yet there is little evidence that progesterone directly affects any of these post-transcriptional events, and, although the hypothesis of transcriptional control is based largely on studies of progesterone action in a non-mammalian model, recent experiments suggest that it may also apply in mammals.

Uteroglobin

Before we can do experiments to determine how progesterone acts in mammalian target organs we need to know the identity of a protein that depends on progesterone stimulation for its synthesis. One such mammalian protein is uteroglobin, a component of endometrial secretion in rabbits.

Uteroglobin is a globular glycoprotein with a molecular weight of 14000 – 15000. It consists of two subunits, each of molecular weight approximately 7000–8000, linked by disulphide bonds, and its isoelectric point is at pH 5.4. Uteroglobin

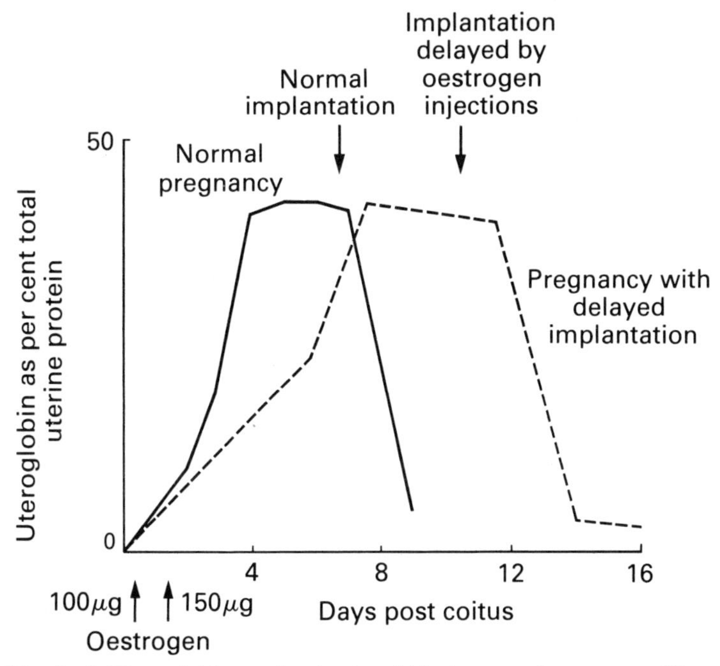

Fig. 6-16. Uteroglobin production in rabbits in normal pregnancy. Two injections of 100 and 150 μg oestradiol benzoate at 6 and 30 hours after mating cause a delay in uteroglobin production. (From H. M. Beier, *J. Reprod. Fert., Suppl.* **25**, 53 (1976).)

secretion is stimulated by progesterone and many synthetic progestagens, but inhibited by oestrogen; it may comprise 50 per cent of the proteins found in the uterine lumen in pregnancy or pseudopregnancy (Fig. 6-16).

The physiological role of uteroglobin remains elusive. Henning Beier in Germany named this unique protein 'uteroglobin', but Joe Daniel in USA called it 'blastokinin' because it allows expansion of rabbit morulae into blastocysts 3 days after ovulation. Recent results from Beier's laboratory show that a normal pattern of uterine secretion is essential for implantation and embryogenesis. Implantation normally occurs on day-$6\frac{1}{2}-7$ after mating. If oestradiol is administered after mating there is a significant delay in the development of the normal pattern of

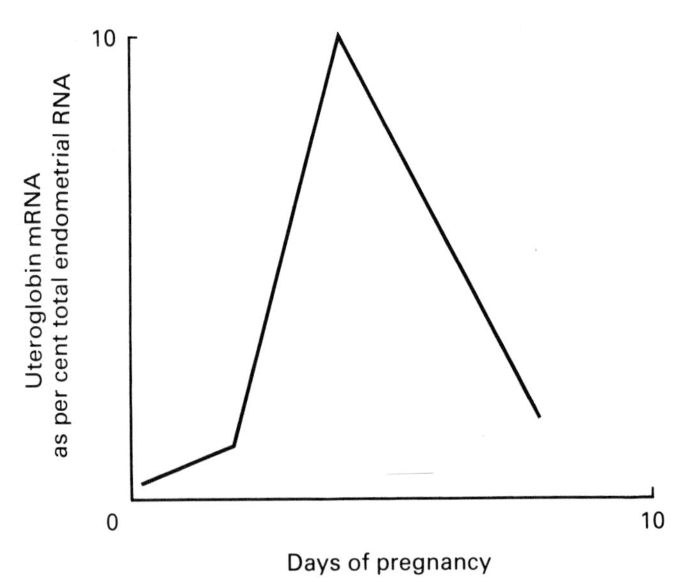

Fig. 6-17. Changes in proportion of uteroglobin mRNA activity in rabbit endometrial RNA during early pregnancy. (From D. W. Bullock, S. L. C. Woo and B. W. O'Malley. *Biol. Reprod.* **15**, 435 (1976).)

uteroglobin production (Fig. 6-16). When normal 4-day blastocysts are transferred to pregnant rabbits treated with oestradiol, implantation is delayed by 4 days, although fetuses will subsequently grow normally and viable young are produced. Conversely, when uteroglobin secretion is advanced by progesterone treatment, 4-day blastocysts will survive in 2-day recipient animals.

Endometrial tissue of rabbits in early pregnancy, or of rabbits treated with oestradiol and progesterone sequentially, contains uteroglobin mRNA in the form of polysomes. Oestradiol alone also causes the accumulation of uteroglobin mRNA in endometrial polysomes, though in much smaller quantities. Dave Bullock working in Bert O'Malley's laboratory has isolated uteroglobin mRNA by techniques similar to those used for avidin mRNA purification. He has shown that the concentration of uteroglobin mRNA in the normal endometrium changes during early pregnancy (Fig. 6-17). Relative to total mRNA, it rises from 0.9 per cent in the non-pregnant endometrium to 10.2 per cent on day 4 of pregnancy and then declines to 1.9 per cent on day 8. Its production therefore parallels that of uteroglobin itself, which reaches its maximum concentration in the uterine lumen of the rabbit on day 4 of pregnancy (Fig. 6-16). After day 7, uteroglobin concentration starts to decline, reaching a low concentration by day 9.

Another property of uteroglobin is its ability to bind progesterone. Work from Miguel Beato's laboratory in Marburg has demonstrated that when the sulphydryl bonds are reduced, uteroglobin has a moderate affinity for progesterone ($K_a = 3 \times 10^6\ \text{M}^{-1}$). If a similar reduction occurs in the uterine lumen, this binding phenomenon may protect the early embryo from potentially high concentrations of progesterone; alternatively intrauterine progesterone may have other regulatory properties in embryonic development, such as acting as a substrate for oestrogen synthesis.

Although the information so far available suggests that the action of progesterone in mammals may be similar to that so far described in chicks, the gaps in our knowledge are clear. Despite its enigmatic role, however, the synthesis of uteroglobin is proving a useful indicator for the biochemical action of progesterone in a mammalian target tissue.

Purple protein

Analysis of uterine flushings from pigs at various stages of the oestrous cycle has led to the identification of a progesterone-dependent protein in endometrial secretions which in some respects is similar to uteroglobin. Uterine flushings contain more protein late in the luteal phase (about 30–40 mg per animal) than at oestrus (< 10 mg per animal), and by polyacrylamide gel electrophoresis, Fuller Bazer and his colleagues at the University of Florida were able to show that at least eight different peptides are induced by progesterone. A major constituent of these induced proteins is a very basic, purple glycoprotein (molecular weight, 32000; isoelectric point at pH 9.7), which imparts the characteristic lavender colour to uterine flushings obtained from progesterone-treated pigs, and accounts for about 15 per cent of all luminal proteins at day 15 of the cycle.

Like uteroglobin in the rabbit, the secretion of purple protein by the endometrial glands of the pig is progesterone-dependent, and large quantities can be obtained from ovariectomized gilts treated with progesterone. Unlike uteroglobin, however, a role in blastocyst development has not been proposed for purple protein. On the basis of its chemical properties, two possible functions have been suggested: first, purple protein binds ferric iron, and may act in some way to transport Fe^{3+} to the developing embryo (this seems particularly plausible in view of an iron-deficiency syndrome found in pregnant pigs, which results in embryonic mortality in mid-gestation); and, secondly, this remarkable protein is also an acid phosphatase, and may function in some unknown way in this capacity *in utero*.

Certainly there is good evidence that purple protein plays some role in embryonic development. Passive immunization of pigs against purple protein results in decreased placental size and allantoic fluid volume; conversely, stimulating uterine production of purple protein by administering high doses of progesterone results in increased placental weight. That purple protein transports something from the endometrium to the conceptus appears likely in view of its appearance in high concentrations in allantoic fluid; it has been shown by immunofluorescence to accumulate in the chorionic areolae, which are probably involved in its uptake and transport into the developing fetus.

Important differences exist between uteroglobin and purple protein. Purple protein itself, unlike uteroglobin, displays no ability to bind steroids. Furthermore, whereas uteroglobin is produced for only 9–14 days after initiation of progesterone treatment, the endometrium of the pig will produce purple protein for at least 60 days under similar conditions. This relationship is consistent with the early effect of uteroglobin on the rabbit blastocyst, compared with the relatively late effect of purple protein, which accumulates in allantoic fluid from about day 30 of pregnancy (i.e. after formation of the chorionic areolae).

Lactose synthetase

An unusual effect of progesterone is its inhibition of lactose synthesis by the mammary gland, which involves the activity of lactose synthetase. In certain species, the enzyme consists of two components and neither is active for lactose synthesis in isolation. One component (a galactosyl transferase) normally catalyses the following reaction:

$$\text{UDP-galactose} + N\text{-acetylglucosamine} \rightarrow \text{UDP} + N\text{-acetyl-lactosamine.}$$

In the presence of the other component, the milk whey protein α-lactalbumin, the K_m for glucose acceptance in the reaction is greatly lowered and the following reaction then occurs:

$$\text{UDP-galactose} + \text{glucose} \rightarrow \text{lactose} + \text{UDP.}$$

The net effect of α-lactalbumin is therefore to change the substrate specificity of galactosyl transferase from N-acetyl-glucosamine to glucose. This finding led to the novel designation of α-lactalbumin as a 'specifier protein'.

The action of progesterone is to regulate the synthesis of α-lactalbumin. During pregnancy when the mammary gland undergoes preparation for milk secretion, the activities of many enzymes increase, including those associated with the synthesis of UDP-galactose. In the mouse, the enzymes involved in the synthesis of N-acetyl-lactosamine also increase during pregnancy, but α-lactalbumin, the lactose synthetase 'specifier' activity, remains at a low level. At, or immediately after parturition, when progesterone levels fall, a rapid rise occurs in the

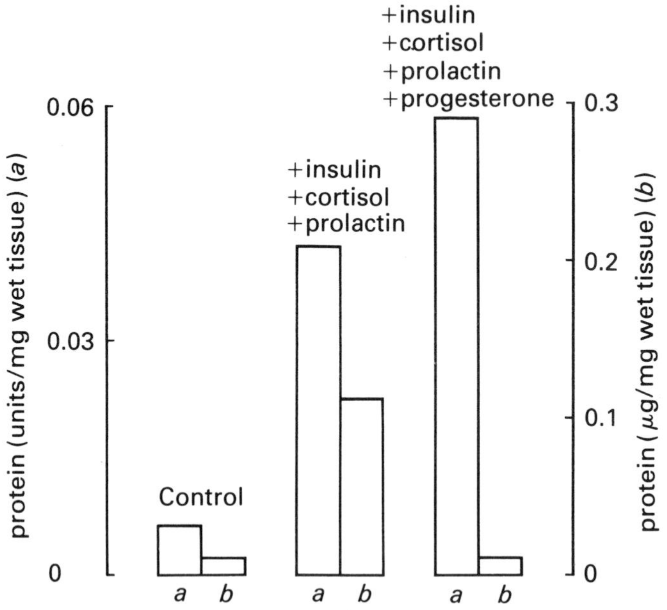

Fig. 6-18. Progesterone inhibition of lactose synthetase (*a*, galactosyltransferase and *b*, α-lactalbumin) in mammary explants from pregnant mice. Explants were grown in tissue culture in the absence of hormones (control), and after the addition of insulin, cortisol and prolactin. The further addition of progesterone (2×10^{-6} M) resulted in a reduction in synthesis of α-lactalbumin. (From R. W. Turkington and R. L. Hill, *Science* **163**, 1458 (1969).)

level of α-lactalbumin in mammary tissue, whereas the galactosyl transferase remains at its pre-lactation level.

The synthesis of whey proteins and casein can be induced in mammary tissue from pregnant mice by incubating the tissue in organ culture with insulin, cortisol and prolactin. Both parts of lactose synthetase are produced in tissue treated in this way but only when treatment includes all three hormones. A striking effect is observed when progesterone is included in the medium. Substantial amounts of galactosyl transferase are synthesized, but the production of α-lactalbumin is inhibited (Fig. 6-18). When pregnant mice are injected with progesterone just before parturition, the increase in α-lactalbumin is also inhibited.

The findings fit the notion that in rats and mice, at least, progesterone prevents a rise in mammary lactose so that when the endogenous secretion of progesterone is reduced experimentally by ovariectomy or naturally at parturition, lactogenesis ensues. This effect of progesterone on lactose synthesis may not be universal since in some species lactose synthesis begins before the time when progesterone concentrations in blood fall.

The idea that in certain target tissues the action of progesterone may be one of inhibition rather than stimulation of protein synthesis is also illustrated by recent work on the induction of casein mRNA. When organ explants of rat mammary gland were cultured with insulin and cortisol, casein mRNA synthesis was induced after prolactin was added to the medium. However, synthesis was inhibited when progesterone was added simultaneously. Although these effects of progesterone on lactose and casein synthesis are not properly understood, Robert Matusik and Jeffrey Rosen of Houston suggest that the hormone acts in a pleiotropic manner to regulate casein synthesis and secretion during pregnancy. There is the growing suspicion that progesterone may inhibit the synthesis of prolactin receptors, prevent the nuclear translocation of nuclear receptors, or directly inhibit casein mRNA transcription.

SYNERGISM AND ANTAGONISM

The physiological effects of progesterone on target cells may be increased or antagonized by oestrogens. One of the best examples of the synergistic effect of oestrogen is the increase it causes in the progestational proliferation of the endometrium in rabbits; this was one of the first progestational effects described in mammals by Pol Bouin more than 60 years ago. Immature or spayed rabbits sensitized with oestrogen produce a greater glandular and epithelial proliferation in response to progesterone than animals treated with progesterone alone. Although this effect has long been regarded as evidence of synergism between the two hormones, the underlying biochemical mechanism has only recently become apparent.

Oestrogen treatment of immature chicks raises the amount of

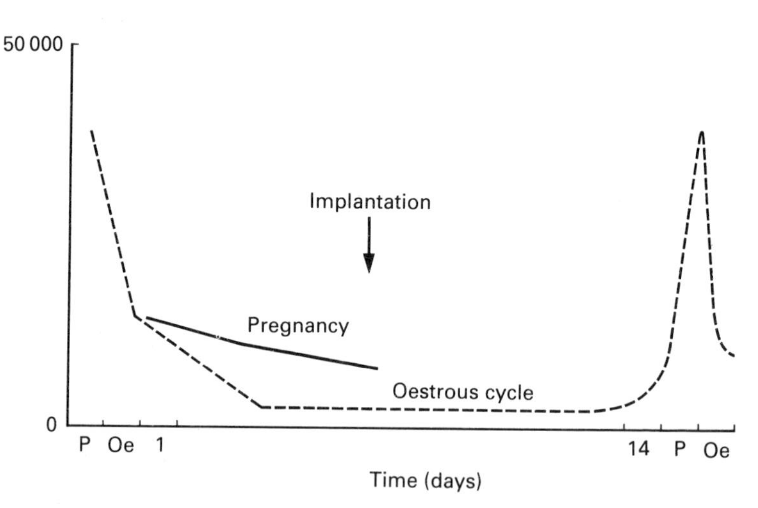

Fig. 6-19. Progesterone-binding sites in the uterine cytosol of cyclic and pregnant guinea pigs. P, pro-oestrus; Oe, oestrus. (From E. Milgrom, M. Atger, M. Perrot and E. E. Baulieu. *Endocrinology* **90**, 1671 (1972).)

progesterone receptor in the oviduct cytosol, but no avidin is produced until progesterone is given. Experiments in guinea pigs and hamsters have shown that oestrogen regulates the synthesis of the progesterone uterine receptor and hence increases the uptake of labelled progesterone by target cells. Edwin Milgrom and his co-workers in Paris, using sedimentation in sucrose density gradients, studied the change in cytosol receptor concentration in the uterus of guinea pigs during the oestrous cycle and after hormone treatment. During pro-oestrus the receptor is predominantly in the form of a 6.7 S component with a smaller proportion of binding in the 4–5 S region; at this stage in the cycle the number of progesterone binding sites per cell, about 40000, is at its highest. In pregnant as in non-pregnant animals the concentration rapidly declines, so that 7 days later at implantation the concentration is low (Fig. 6-19). Oestrogen stimulates the production of progesterone cytosol receptors of

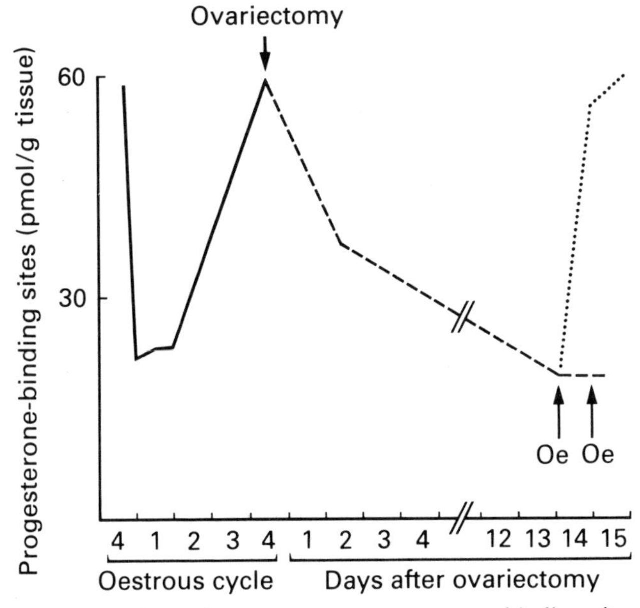

Fig. 6-20. Effect of oestrogen on progesterone binding sites in the uterine cytosol of hamsters. Concentration of binding sites during the oestrous cycle is shown by a solid line. The effect of ovariectomy is shown by the dashed line which reveals a slower decrease than that found after pro-oestrus. The effect of oestrogen treatment (15 μg per day oestradiol, Oe) is shown by the dotted line. (From W. W. Leavitt, D. O. Toft, C. A. Strott and B. W. O'Malley. *Endocrinology* **94**, 1041 (1974).)

the 6.7 S form, their number increasing from a few thousand per cell to about 30000 sites per cell in about 24 hours. Receptor synthesis is inhibited by actinomycin D or cycloheximide – so it depends on the synthesis of both RNA and protein.

A similar control mechanism operates in hamsters. The concentration of progesterone-binding sites is closely correlated with endogenous oestradiol values in blood and reaches its highest concentration on the day of pro-oestrus; ovariectomy at this time causes a fall in the number of progesterone binding

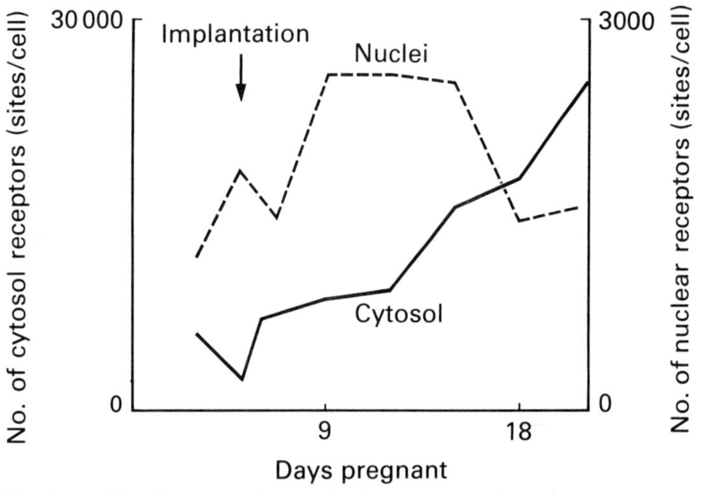

Fig. 6-21. Simultaneous changes in the concentration of progesterone–receptor complexes in uterine cytosol and nucleus of the pregnant rat. Note the decrease in cytosol binding, and the increase in nuclear binding about the time of implantation (Day 5–6). (From E. Milgrom *et al. Endocrinology*, vol. 1, p. 478. Ed. V. H. T. James. Amsterdam; Excerpta Medica. (1977).)

sites, whereas oestradiol treatment 14 days later restores them to their pro-oestrous value (Fig. 6-20).

These findings imply that the synergism observed between progesterone and oestradiol is related to the synthesis of progesterone receptors during oestrogen sensitization. But at first sight it seems paradoxical that in the guinea pig and rat the progesterone–receptor concentration in the cytosol reaches its lowest values at the time of implantation. However, we have so far considered only the cytosol receptors, and the explanation may lie in a rise in number of receptors in the nucleus (Fig. 6-21). Edwin Milgrom suggests that this nuclear increase could be the result of translocation under the influence of increasing circulating levels of progesterone, and that such a mechanism elicits the physiological changes necessary for implantation to proceed.

Another form of interaction is the control of the oestrogen receptor by progesterone. Progesterone has long been known to antagonize the effects of oestrogen in the chick oviduct, rat uterus and rabbit endometrium; it will inhibit or reduce the effect of oestrogen on oestrogen-induced growth of these tissues, though neither the initial binding of oestrogens, nor their translocation into the nucleus are prevented. The site of antagonism must therefore reside elsewhere.

James Clark and Ernest Peck have studied this problem by investigating the synthesis and intracellular recycling of cytosol oestrogen receptor in the rat uterus, using procedures designed to measure the total amounts of cytosol and nuclear receptors. They have demonstrated that, when oestrogen-sensitized immature rats are given an oestradiol injection, the quantity of cytosol receptor declines as would be expected from translocation, and then increases 4–8 hours later as the receptor is replenished, possibly by recycling from the nucleus. A second phase then occurs, 8–24 hours after injection, when the receptor concentration increases, probably by synthesis. This phase can be blocked by progesterone if it is injected at the same time as oestradiol. Etienne Baulieu and his coworkers showed that the first phase (recycling) could not be blocked by a protein inhibitor, cycloheximide, whereas the second phase (synthesis) depended on protein synthesis.

The above results show that the synergistic and antagonistic effects of oestrogen and progesterone in target cells can now be viewed in biochemical terms and the mechanisms involved have been summarized by James Clark and his colleagues (Fig. 6-22). Oestrogen sensitization stimulates the production of a progesterone receptor in the cytosol of target cells. Progesterone binds to this receptor and is translocated to the nucleus where it decreases the synthesis of oestrogen receptor, and reduces the retention time of the oestrogen receptor in the nucleus. In this way progesterone redirects the cellular response from the tissue hypertrophy and hyperplasia associated with oestrogens to the progestational secretory activity that prepares the uterus for implantation and the maintenance of pregnancy. An important caveat should be added at this point, however, since most of the above conclusions have been derived from studies on total

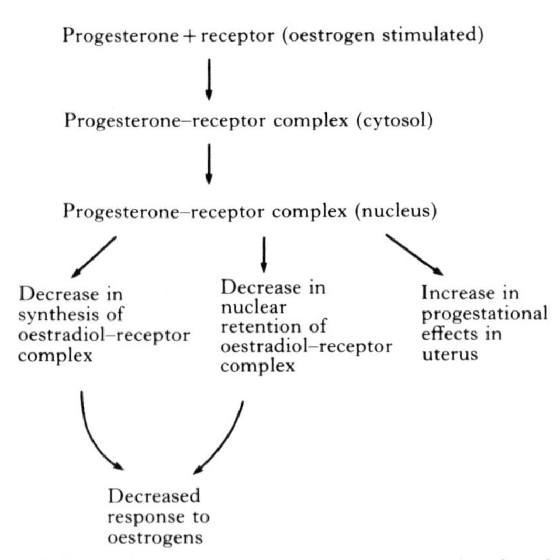

Fig. 6-22. Effect of progesterone on oestrogen-stimulated uterus; a scheme to describe the sequence of biochemical events. (From J. H. Clark, E. J. Peck and S. R. Glasser. *Reproduction in Domestic Animals*, 3rd edn Ed. H. H. Cole and P. T. Cupps, p. 143. New York; Academic Press (1977).)

uterine homogenates. Current experiments indicate that significant differences may exist between various cell types in their response to oestrogen stimulation of progesterone receptors. The development of more stringent techniques to separate epithelial, stromal and myometrial cells should eventually clarify whether the scheme depicted in Fig. 6-22 applies only to certain cell types within the uterus.

Whereas the properties of progesterone and oestrogen normally stand in marked contrast, a final bizarre example will illustrate that in some circumstances the two hormones can substitute for each other and each may even mimic the other's molecular action. This remarkable situation is to be found in the stimulation of ovalbumin synthesis in the chick oviduct; as yet, no parallel example is known in mammals. Ovalbumin secretion by the untreated chick oviduct is stimulated by oestrogen, and progesterone is entirely without effect (see Chapter 5). However, 12 days after oestrogen is withdrawn, progesterone will stimulate ovalbumin secretion. In this secondary stimulation, progesterone mimics precisely the action of oestrogen causing a rapid increase in the number of available RNA polymerase sites. Both hormones enhance transcription at similar chromatin sites, and induce the accumulation of ovalbumin mRNA in oviduct cells. Moreover, when chromatin is prepared from chicks in which oestrogen treatment had ceased, synthesis of RNA in a cell-free system results in a ten-fold increase in ovalbumin mRNA production in the presence of the progesterone–receptor complex.

Fascinating parallels exist between the ways in which different steroid hormones act on target cells. They all bind to specific receptors in the cytosol and are translocated to the nucleus, and they all affect the transcription of the genome, inducing proteins characteristic of individual organs. Although in the present chapter we have described new findings about the early effects of progesterone, there is still much to be learned about how these primary events are translated into hormone-dependent phenomena such as implantation, the inhibition of myometrial activity, and the feedback control of hypothalamo–pituitary secretion.

The new findings described in this chapter tell us about the molecular action of steroid hormones and also something relevant to the evolution of hormones. Some organs react to steroids because they have specific receptors in the cytosol, and acceptor sites in the nuclear chromatin which bind the steroid–receptor complexes to the genome. In 1953, Sir Peter Medawar alluded to the economy of evolution in utilizing derivatives of the steroid nucleus – a chemical form widely distributed in nature – as regulators of biological processes. He referred to this as 'endocrine evolution' and described it 'not as an evolution of hormones but an evolution of the use to which they are put'. In the last 20 years we have begun to understand this basic biological truth in chemical terms.

SUGGESTED FURTHER READING

The receptors of steroid hormones. B. W. O'Malley and W. T. Schrader. *Scientific American* **234**, 32 (1976).

Female steroid hormones and target cell nuclei. B. W. O'Malley and A. R. Means. *Science* **183**, 610 (1974).

The mechanism of action of estrogens and progesterone. R. B. Heap and D. V. Illingworth. In *The Ovary*, vol. 3, pp. 59–150. Ed. Lord Zuckerman and B. J. Weir. New York; Academic Press. (1977).

Mechanisms of action of sex steroid hormones in the female. J. H. Clark, E. J. Peck and S. R. Glasser. In *Reproduction in Domestic Animals*, 3rd edn. Ed. H. H. Cole and P. T. Cupps, pp. 143–73. London; Academic Press. (1977).

A review of regulation of gene expression by steroid hormone receptors. B. W. O'Malley, R. J. Schwartz and W. T. Schrader. *Journal of Steroid Biochemistry* **7**, 1151 (1976).

Biochemical actions of progesterone and progestins. Ed. E. Gurpide. *Annals of the New York Academy of Sciences* **286** (1977).

Index

Index